Maria E. Catlow

Popular British Entomology

Containing a Familiar and Technical Description of the Insects...

Maria E. Catlow

Popular British Entomology
Containing a Familiar and Technical Description of the Insects...

ISBN/EAN: 9783744643603

Printed in Europe, USA, Canada, Australia, Japan

Cover: Foto ©berggeist007 / pixelio.de

More available books at **www.hansebooks.com**

POPULAR

BRITISH ENTOMOLOGY;

CONTAINING

A FAMILIAR AND TECHNICAL DESCRIPTION OF THE INSECTS
MOST COMMON TO THE LOCALITIES OF THE

BRITISH ISLES.

BY

MARIA E. CATLOW.

𝔄 𝔑𝔢𝔴 𝔈𝔡𝔦𝔱𝔦𝔬𝔫.

LONDON:
ROUTLEDGE, WARNE, & ROUTLEDGE,
FARRINGDON STREET;
NEW YORK: 56, WALKER STREET.
1860.

TO

A. & S. J. PIGGOTT,

Of Fitzhall, Sussex,

THIS LITTLE WORK,

INTENDED TO FACILITATE AND ENCOURAGE THE PURSUIT

OF

British Entomology,

IS MOST AFFECTIONATELY INSCRIBED,

BY THEIR SINCERE FRIEND,

THE AUTHOR.

BRIDGELAND,
August 21st, 1848.

PREFACE.

THE deeply interesting branch of Natural History which this little volume is intended to illustrate is one that cannot fail to prove attractive to those who engage in its pursuit; not only from the extreme beauty and variety of colour and form displayed by the insect world (though in these respects almost unrivalled), but also from the unbounded proofs of the Divine goodness and wisdom evinced, in the adaptation of each member in its numerous families to the purposes of creation; and in the instincts bestowed upon them, by which these purposes are fulfilled.

The limits assigned to the work will, of course, only admit the description of a portion of the species indigenous in this country; but it is hoped that an acquaintance with those insects which have been selected as examples of their class, for their superior beauty or interest, will prepare the student for the profitable reading of those extended works, which he will more readily comprehend, from having his attention directed, in the commencement of the study, to a comparatively few leading particulars.

When these are well understood, and a tolerably accurate idea is gained of the peculiarities of the Orders and Families, the young Entomologist will turn with more satisfaction and interest to the examination of works devoted to the elucidation of the Genera and Species. For this purpose the advanced student should consult Curtis's "British Entomology," which has the advantage of beautifully-executed coloured illustrations; Westwood's "Entomologist's Compendium" and "Introduction to the Classification of Insects;" and for ample information on the history, habits, and uses of the insect tribe, Kirby and Spence's Introduction will be found highly profitable and interesting.

The attention of the reader may also be profitably directed to another work entitled "Episodes of Insect Life," in which the Instincts and Habits of Insects are described through the agreeable medium of a series of Essays, combining in a novel form an admixture of the Real and Ideal of Entomological life, adapted to the months of the year.

To the works of Curtis, Westwood, and Kirby and Spence, the author of this volume is deeply indebted, and their valuable details being too voluminous and expensive for the generality of young students, will, she trusts, be deemed sufficient apology for the endeavour to open an easy and pleasant path to that pursuit, which they have so ably illustrated by their talents and research.

In the first chapter, the author has given a concise introduction to the subject, explanatory of the different parts into which insects are scientifically divided by Entomologists, omitting those minor details only, which are merely useful to the anatomist, or to those wishing to enter more minutely into the dissection of specimens, than the readers

of this little work are supposed able or willing to do.   The
second chapter is devoted to the classification ; the eight
following embrace a description of as many species as the
space will admit, with notices of the most interesting
families, their uses and peculiarities ; and the two last
chapters contain an account of the various methods of
capturing and preparing insects for the cabinet; the in-
juries and benefits derived from them by man ; a short
account of fossil insects, abridged from Burmeister's Introduction ; with other useful and interesting particulars.

LIST OF PLATES.

POPULAR

BRITISH ENTOMOLOGY.

CHAPTER I.

JANUARY.

THERE is no branch of Natural History more interesting, and few more within the reach of students, than that of Entomology : none, certainly, in which they will find more abundant proofs of the wisdom and goodness of that great Being, who is equally the Creator of the gigantic whale and the diminutive animacule,—of the majestic oak which flourishes a thousand years, and the fleeting ephemera which fulfils its appointed duties, and lays down its little life in the space of a few hours,—of the mightiest intellect which ever adorned the world, and the instinct which teaches the parent moth to lay its eggs on that plant which will afford fitting

B

nourishment to the young larvæ so different to herself in form and habits.

The insect world is replete with wonders, and a whole life passed in research would fail to exhaust the pleasure derivable from its attentive pursuit, more particularly to those who in the study of the creature do not lose sight of the Creator, but look

"Through Nature up to Nature's God."

What, for instance, can be more beautiful or surprising than the transformations which most insects undergo ? Hatched in the form of a caterpillar, then becoming nearly torpid in the pupa state, which they exchange for that of the imago or perfect insect, and in each of these very different conditions, exhibiting instincts adapted to the medium in which they are placed, and the functions necessary to their existence. A little insight into the subject will be gained by the following explanation :—The four stages of an insect's life are the egg, larva, pupa, and imago. The eggs are usually of an oval form, though some species vary considerably in shape, being globular, conical, cylindrical, pear-shaped, etc. They are for the most part smooth, but many are beautifully ornamented, the colour also varying considerably : white, yellow, and green are the usual tints, yet orange, red, brown, black, and blue are to be found ; others are banded, as, for example, those of the Pine Moth, which are blue, with three brown zones. The number of eggs deposited by insects is very considerable : the Queen Bee produces from 40,000 to 50,000 in the course of the year ; one species of Moth, according to Lyonnet, produces 1,000,000 young in the third generation, and the Aphis observed by Réaumur produced at the fourth generation 5,904,900,000 individuals. The period which ensues before the egg is hatched, depends on the temperature of the atmosphere, as well as the kind of insect, varying from a few hours, as in the Meat-fly, to some months, and as a general rule those eggs which are laid in the autumn are not hatched before the next spring. In the larva, the next state of existence, the principal occupation of insects is to eat and grow : in this condition they are called caterpillars, grubs, or maggots, but the proper and more scientific name is *larvæ*. Of these there are two kinds—those which in their general form have some resemblance to the perfect insect, such as the Orders *Orthoptera, Hemiptera, Homoptera,* and part of the order of *Neuroptera ;* and those which are quite unlike, comprising, with few exceptions, the Orders *Coleoptera, Lepidoptera, Hymenoptera, Diptera,* and part of *Neuroptera.* These last-mentioned larvæ are of various forms, and are divided into two principal classes—those having a distinct scaly head, including many of our principal insects, and those without a distinct head, as the majority of the two-winged Flies ; the minor distinctions depend principally on the number of the legs and prolegs, but for these the student must refer to larger works. Having attained its full size after several times changing the skin, the larva undergoes its final and most important transformation, appearing in quite a different shape, that of the *pupa*, during the period of which it remains more or less inactive, and would be liable to destruction were it not for the admirable instinct shown by the larva in preparing for the new state of existence to which it is destined. Here again the distinctions are too multiplied to be fully entered into in so slight a sketch as the present, though the whole may be classed under two heads—those whose transformation is partial, and those in which it is complete. Linnæus divided them into five kinds, as follows :—

Pupa completa, active, with all the parts of the perfect insect : example, *Aranea*, the Spider.

Pupa semi-completa, active, resembling the parent, but having only the rudiments of wings : example, *Gryllus*, the Grasshopper.

Pupa incompleta, inactive, but with rudimental legs and wings : example, *Apis*, the Bee.

Pupa obtecta, the thorax and body distinct, enclosed in a scaly covering, and either naked or in a cocoon : example, *Lepidoptera*, the Butterfly and Moth.

Pupa coarctata, enclosed within a cage formed of the skin of the larva : example, *Musca*, the Fly.

The first of these is now inadmissible, as the creatures composing it are no longer classed with insects. More modern entomologists have altered these arrangements, particularly Latreille ; but leaving these somewhat intricate questions, we will endeavour rather to gain a general idea of this state of the insect world, which is interesting from its being prior to the imago or perfect condition.

Pupæ are generally of a dirty white colour, when their habitation is underground ; some however are of a dark bright brown, and those which have more exposure to the action of light are variable in their tints, some being of bright colours and gilded, as in the chrysalides of many Butterflies. The period passed in the pupa state depends greatly on the temperature of the atmosphere, as Providence has wisely ordained that the development of the perfect insect shall not take place till its proper sustenance

is prepared, and a suitable provision made for the deposi-
tion of the eggs in a proper locality.

The *imago*, or perfect insect, on first emerging from the
pupa-case, is frequently in an imperfect condition.   In the
Order *Lepidoptera*, for instance, the antennæ are bent
down, and the wings crumpled, small, and shapeless ; but
in a short time these are gradually unfolded, and assume
their proper form ; the elytra of the *Coleoptera* gain their
beautiful colours, and what before seemed a half-formed
mass is changed into an insect decked with the most brilliant hues, and rejoicing in its new and happy existence.
The operation of expanding the wings generally occupies
but a few minutes, though some *Butterflies* require nearly
an hour, and several species of *Sphinx* even a day.   In the
Ephemera this process is almost instantaneous, some of them
however undergoing another slight metamorphosis after
they have quitted the puparium : fixing themselves upon
some object, they draw every part of the body, even the legs
and wings, from a thin pellicle or skin, which covered them
like a glove, and so exactly do these exuviæ resemble the
insect, that they may at first sight be mistaken for it.   Many
similar examples might be brought forward, as proofs of the
interesting facts discoverable in the study of these " living

gems," but as they will be better understood and more appreciated when the technical terms are a little more familiar, we will proceed to the explanation of those more commonly used, after giving a slight sketch of the plan pursued in this brief epitome.

As this work is intended for those who wish to gain a somewhat scientific insight into the subject, yet are wanting in time or inclination to dive into the many valuable works written on this branch of Natural History, I shall first give a short account of the Classes and Orders established by those modern Entomologists who are deemed good authorities (though every year, in the present advancing state of the science, calls for slight changes and modifications in the system) ; and then commencing with the month of March, the first in which the study would be prosecuted with much success, describe as many of those insects as the limits of the work will allow, which are likely to attract the eye or interest the mind of the young student. These will embrace the whole of the British Butterflies, as being the most beautiful and attractive of the insect world ; many of our most common Moths and Beetles; and such other insects as are likely to come under the observation of the lover of Nature. To those who are sufficiently interested in the pursuit to devote to it more time and research, the valuable works of Kirby and Spence, and many others, will afford ample occupation and enjoyment ; and should any be led to these by this unpretending little pilot, the writer will proudly resign them into the hands of more able guides, satisfied that they cannot examine even the eye of an insect, or the down on it wing, without strengthening their belief in the power, wisdom, and goodness of the Creator.

The branch of Natural History we are about to examine derives its name, Entomology, from two Greek words, signifying *an insect* and *a discourse ;* the term insect, from the Latin *insecta,* meaning an animal divided into numerous parts or segments.   Insects are further distinguished by being invariably furnished with six legs, a head distinct from the body, two antennæ, and by their breathing through pores, or tracheæ.

Insects are divided into three parts : the *head,* which contains the organs of sensation; the *thorax,* or trunk, which includes the organs of motion ; and the rest of the body, containing the organs of respiration.

With respect to the *mouth,* insects have been divided by some authors into two sub-classes,—MANDIBULATA, those furnished with two pair of jaws, called *mandibulæ* and *maxillæ,* by which they gnaw their food; and HAUSTELLATA, where these jaws are replaced by minute laminæ forming a kind of sucker, *haustellum,* which is received into a sheath, —the insects of this class procure their food by suction, either from animal or vegetable substances. The other parts of the mouth are the *palpi,* or feelers, small movable appendages placed on each side, which vary in size and number; the *promuscis* or *rostrum* being the part forming the mouth in many of the sucking insects; the *proboscis,* or sheath, which contains the *trophi,* or organs of the mouth in Dipterous insects; and the *antlia,* or suckers, the organs constituting the principal part of the mouth in *Lepidoptera.* This organ is generally very slender; in the family *Sphingidæ* it is long, in *Papilionidæ* much shorter; in a state of rest it is rolled spirally between the palpi.

The *eyes* of insects are of two kinds, simple and compound, horny, immovable, and unprotected by any eyelid. When closely examined, they are found to consist of a great number of minute hexagonal lenses, each of which forms a distinct organ of vision; of these, the common Fly possesses 4,000, the Silkworm Moth 6,236, and some Butterflies 17,355: when detached from the head and cleaned, they are found to be as clear as crystal. The celebrated natu-

ralist Réaumur fitted one to a lens, and found that he could
see through it distinctly, the object being greatly magnified.
Sometimes the eyes are divided by means of the antennæ,
which are inserted in the middle, but generally they are
entire ; in some instances they are placed at the end of
footstalks, and in a few insects the eyes seem entirely
wanting.   The organ of hearing is not manifest in insects,
though most of them possess this faculty in a certain degree,
as is evident from the power many individuals have of pro-
ducing sound, which would be a useless gift were the other
wanting.   The sense of smell is shown by their instantly
discovering and crowding to those places where food agreeable to them is to be found.   Some naturalists are inclined
to believe this sense to exist in the antennæ ; others transfer
the sensation to the palpi, and from experiments which have
been made on Bees, it seems probable that the chief sensa-
tions are communicated by the mouth, from the fact of the
proboscis being more or less developed, as the palpi are
minute or wanting.   M. Lamarck considers this idea as
probably the correct one.

The organs of touch are usually supposed to exist in the
antennæ, or feelers ; for the body, being in general hard or
horny, can but slightly communicate the sense of feeling.

The antennæ are the two movable appendages, like horns, which are so characteristic of insects ; they are of various forms, and assist greatly in determining the genera ; it is therefore necessary that the student should be acquainted with the principal varieties they assume, and the names by which they are distinguished. Some authors present a formidable list, but the most common are as follows:—*Seta-ceous*, when they gradually taper towards the extremity. *Clavate*, when they grow thicker from the base. *Filiform*, when of an equal thickness throughout. *Moniliform*, when composed of a series of knots resembling a string of beads. *Capitate*, when they terminate in a knob. *Fissile* when the knob is divided longitudinally, into laminæ or plates. *Per-foliate*, having the knob divided horizontally. *Pectinate*, having a longitudinal series of processes like hairs projecting from them in the form of a comb. *Plumose*, when they resemble a feather. *Prismatic*, when like a prism, or formed of three sides. *Fusiform*, small at the two extremities and thick in the centre, like a spindle. *Furcate*, when the antennæ are divided into two branches like a fork.

The number of joints in the antennæ varies considerably. Coleopterous insects have in general eleven joints ; the stinging portion of the order Hymenoptera twelve or thir-

teen, according to the sex; in *Lepidoptera* and some others
they are much more numerous. A great diversity also
often exists in the structure of these organs in the opposite
sexes of the same species.

The *elytra*, or wing-cases, are the hard coverings which
conceal the wings of the order *Coleoptera* and others; they
open longitudinally, and the difference in their form affords
generic and specific characters. The *alæ*, or wings, are the
organs of flight: insects possess either two or four, when
not entirely wanting; where there are only two, they are
of a uniform size and appearance; when four, they most
frequently differ, the first or anterior pair being larger than
the other. These appendages are membranous, elastic, generally transparent; attached to the upper side of the thorax,
and intersected with nerves, which sometimes form a kind
of network. In Wasps and Bees the wings are naked and
transparent; in Butterflies they are covered with minute
scales, embellished with the liveliest colours; in the Cad-
dice Flies they are clothed with fine hairs, whence the name
of the order *Trichoptera*. *Halteres*, or poisers, are two
short movable appendages placed near the origin of each
wing; this organ is peculiar to the two-winged insects.

The *aculeus*, or sting, is the instrument by which insects

wound and instil a poison ; it is used both as an offensive and defensive weapon. The *ovipositor* is the instrument which serves to pierce wood or the bodies of animals, in order to deposit eggs.

The *legs* in insects are divided into the *coxa*, or hip, of two joints ; the *femur*, or thigh ; the *tibia*, or leg ; and the *tarsus*, or toe. The number of joints in the tarsus is in some orders constantly five, but in others it varies from one to five, and sometimes the posterior or hinder tarsi have a joint less than the anterior. Upon the differences in these numbers are established the chief divisions in the order *Coleoptera*. The last joint of the tarsi is almost always terminated by two hooks.

In the form of the feet and more particularly of the *tarsi*, there are many varieties, according to the habits of the insects. The anterior pair have sometimes the *femur* or thigh grooved and armed with dentations, and the tarsi terminated by a strong spine : insects having the anterior feet constructed in this manner, use them for seizing their prey, and are termed *Raptorii*. Others have the tarsi flat and hairy, this form enabling the insect to employ them as oars, or for swimming ; these are called *Natatorii*. In the Bee family the legs are formed in such a manner as to

brush off the pollen from the stamens of flowers ; and in
other species the anterior legs are broad and spined, thus
being calculated for digging in the earth.   When the upper
part of the leg is slender and cylindrical, the motion is
generally confined to walking ; when thick and apparently
swelled, the insect is usually capable of swimming or leaping,
as this peculiarity indicates greater muscular power.   The
muscles in most insects are very numerous, and of course
exceedingly minute ; in the Caterpillar of one species,
Lyonnet counted more than four thousand, while those of
the human body do not exceed five hundred and twenty-
nine ; from this cause many Caterpillars can suspend themselves in a horizontal posture for some hours ; most persons
have noticed one curious species which has the appearance
of a dried twig, both in colour and in the perfect rigidity of
the attitude it assumes.

These are the principal terms necessarily used in the
description of insects ; others more technical may be dispensed with till the student is further advanced in the
science.   The next month may be occupied in studying
the modern classification, adopted with some modifications
by Leach, Latreille, Kirby, Curtis, Westwood, and other
modern Entomology.

CHAPTER II.

FEBRUARY.

WE will now enter briefly into an explanation of the
Orders which this little work is intended to illustrate, pre-
mising that in consulting other works when advanced in
the study, the young entomologist must expect to meet with
various opinions on these, as well as the minor divisions
in the classification. Orders are entirely omitted by some
which are considered indispensable by others, or new names
introduced where the order itself is recognized. This diversity of arrangement cannot be wondered at, when we
consider the vast number of insects to be classed, the great
variety in their structure, and the constant discoveries
which are made in the science. The system here adopted
is that followed by many eminent naturalists, and some

orders are retained, though considered unnecessary by other entomologists of equal celebrity ; these may be easily dismissed by those students who gain sufficient experience from the works of Nature and the study of science to be enabled to form their own opinion ; while to the mere amateur it is of little importance, and certainly unadvisable, to enter into questions which the learned cannot satisfactorily set at rest.

Insects constitute the fifth class in the third sub-kingdom of animated nature. Some authors make two great divisions, *A metabola,* insects undergoing no metamorphosis, and *Metabola,* those which undergo the three great changes from larva to pupa, and thence to the imago or perfect insect ; but the latter of these divisions only will be considered, as the former consists of insects little interesting to a beginner in the science, and indeed their situation is much disputed by entomologists. They consist of THYSANAURA, divided into two small families, *Lepismidæ* and *Poduridæ,* minute insects without wings, found under stones; and floating on water ; and ANOPLURA, also divided into two families, *Pediculidæ* and *Nirmidæ,*—these consist of the parasitic insects which are found living on animals and birds, known by the general name of Lice.

METABOLA,

Insects undergoing transformation.

ORDER I. COLEOPTERA, *Aristotle*.

Characters.—*Elytra* horny or crustaceous, with a straight suture or opening; *wings* two, membranous, longitudinally and transversely folded when at rest; *mouth* with transversely movable jaws; *antennæ* generally consisting of eleven joints.

This Order is divided into four sections and many families.

SECTION I.

PENTAMERA.—Tarsi five-jointed.

1. *Cicindelidæ.*—Antennæ long and slender; legs long and formed for running; maxillæ armed with a claw; body generally oblong; eyes prominent; colour generally brilliant; size moderate. Locality, hot sandy districts.

2. *Carabidæ.*—Antennæ filiform; legs long and slender; maxillæ with no movable claw; body oblong; eyes but slightly prominent; wings often wanting. Locality, under stones and moss.

3. *Dyticidæ.*—Antennæ long and slender; anterior legs short; hinder pair formed for swimming; mandibles short and strong; maxillæ curved from the base; body

c

oval, and generally depressed ; eyes slightly prominent ; colour generally dark brown, olive, or black. Locality, stagnant water, but occasionally found flying.

4. *Gyrinidœ.*—Antennæ very short ; anterior legs long ; four posterior legs very short and flat ; mandibles short, horny, and notched ; maxillæ flat, curved, and acute at the tip ; body generally ovate and depressed ; eyes consisting of two pair; colour metallic ; size small. Locality, running water, though sometimes found flying.

5. *Hydrophilidœ.*—Antennæ clavate; legs formed for swimming, and hairy ; mandibles much toothed ; maxillæ, outer lobe short and broad ; body hemispherical and convex ; maxillary palpi very long; colour generally obscure ; size large. Locality, water and water-plants.

6. *Silphidœ.*—Antennæ thickened at the tip ; mandibles very strong ; maxillæ with two lobes ; palpi filiform ; body depressed ; colour obscure ; size moderate. Locality, dead animal matter.

7. *Dermestidœ.*—Antennæ short and clavate ; legs short; jaws short and thick ; body thick and oblong, rounded at each end, and clothed with scales or hairs, which give it a variety of tints ; size small. Localities, dried skins and fur, also bacon, books, etc.

8. *Staphylinidæ.*—Antennæ rather short ; legs robust ; jaws very powerful ; body long, narrow, and depressed in form ; elytra much shorter than the body ; colour frequently black. Localities, putrescent plants, fungi, flowers, and under the bark of trees.

9. *Byrrhidæ.*—Antennæ more or less clavate ; legs contractile, that is, the tarsi can be laid close to the surface of the tibia ; elytra covering the body ; body short, oval, or rounded, very convex ; colour obscure. Localities, sand-pits and foot-paths.

10. *Histeridæ.*—Antennæ short and elbowed; legs toothed, the two posterior pair being inserted widely apart ; mandibles very robust ; maxillæ long ; palpi filiform ; elytra short and truncate, leaving the body exposed ; body square and highly polished ; colour black and shining, sometimes varied with spots ; size small. Localities, bark of trees and dung.

11. *Lucanidæ.*—Antennæ elbowed, and terminated by a club ; fore legs generally longer than the others ; mandibles very large ; elytra covering the body ; body oblong, oval, and depressed ; colour black ; size large. Locality, trunks of trees.

12. *Geotrupidæ.*—Antennæ with a large club; mandibles

horny and curved ; elytra smooth or simply striated ;
body short, thick, and convex ; colour metallic underneath, generally black above ; size moderate. Locality, fresh dung.

13. *Scarabidæ.*—Antennæ with a three-jointed club ; legs
long, posterior pair placed far back ; clypeus * large
and advanced ; mandibles terminating in a long membranaceous plate ; elytra somewhat square behind ;
body broad and depressed ; colour usually black ; size
small. Locality, manure.

14. *Aphodidæ.*—Antennæ with a three-jointed club ; mandibles short and dilated ; legs at equal distances ; body
elongated, oblong, or oval, and rounded at the extremity ; clypeus large, concealing the mandibles ; colour
often black ; size very minute. Locality, sandy situations.

15. *Trogidæ.*—Antennæ with a three-jointed club ; mandibles horny ; maxillæ with two lobes ; head deflexed ; †
wings sometimes wanting ; body ovate ; colour obscure.
Locality, sand-pits.

16. *Melolonthidæ.*—Antennæ with a large club, of several
plates ; mandibles robust and horny ; maxillæ with acute

* Shield of the mouth.   † Bent down.

teeth ; clypeus large ; elytra shorter than the body ; body ovate and convex ; colour rarely metallic, the body being generally covered with minute scales; size large. Localities, trees and hedges.

17. *Cetonidæ.*—Antennæ ten-jointed ; mandibles slender and compressed ; maxillæ horny, with the inner margin membranous ; elytra shorter than the body ; body depressed, generally oval, the front of the breast sometimes produced to a point ; colour brilliant ; size moderate. Localities, trees and flowers.

18. *Buprestidæ.*—Antennæ short and serrated ; legs short ; mandibles short, entire at the tips, and triangular ; elytra narrow at the tips ; body oblong and depressed, sometimes linear, with a projection in front in some species, in others, at the end of the body ; colour very splendid in exotic species ; size small. Localities, forests and timber.

19. *Elateridæ.*—Antennæ short and serrated ; legs short ; mandibles bifid ;* maxillæ two-lobed ; body long and narrow, produced behind into an acute point ; colour obscure ; size small. Localities, flowers, plants, and grass.

* Divided into two at the end.

20. *Lampyridæ.*—Antennæ filiform; mandibles small, acute, and curved ; maxillæ small ; elytra sometimes short ; eyes large ; body long, depressed, and soft ; colour obscure ; size small. Localities, meadows and hedges.

21. *Telephoridæ.*—Antennæ simple and moderately long ; mandibles acute and curved ; wings and elytra perfect; body long, narrow, and soft; colour generally dull; size small. Locality, flowers.

22. *Melyridæ.*—Antennæ moderately long, serrated or pectinated ; body long, ovate, slightly convex and soft ; colour bright green and red, conspicuous ; size small. Locality, flowers.

23. *Cleridæ.*—Antennæ short, sometimes filiform, and serrated, sometimes clavated ; mandibles with several teeth ; body long, often cylindric ; colour variegated ; size small. Localities, flowers, bark, and dry wood.

24. *Ptinidæ.*—Antennæ moderately long, filiform, pectinate or serrated ; mandibles small and toothed ; body oval, but generally obtuse at each end; colour obscure. Localities, old houses, paling, and stumps of trees.

25. *Lymexylonidæ.*—Antennæ short ; mandibles short, thick, and obtusely toothed ; head separated from the thorax by a narrow neck ; body linear. Locality, woods.

26. *Bostrichidæ.*—Antennæ short, with a three-jointed club; mandibles robust ; thorax forming a shield over the head ; body cylindrical ; size small. Locality, trunks of trees.

27. *Scydmenidæ.*—Antennæ as long as the head and thorax; mandibles toothed ; legs long and slender ; size very minute. Locality, under stones and moss.

SECTION II.

HETEROMERA.—Five joints in the first four tarsi, and four in the hindmost.

1. *Notoxidæ.*—Antennæ filiform, or slightly thickened at the tips ; mandibles strong, triangular or quadrate ; maxillary palpi terminated in a large hatchet-shaped joint ; thorax narrow behind ; size minute. Localities, sandy situations and various plants.

2. *Pyrochroidæ.*—Antennæ in the males generally pectinated or serrated ; mandibles acutely notched at the tips ; maxillary palpi serrated ; body narrow in front; elytra covering the body ; colour red. Localities, flowers and leaves.

3. *Lagriidæ.*—Antennæ filiform ; mandibles short and thick ; maxillary palpi, last joint triangular ; head and

thorax narrower than the elytra.   Localities, hedges
and woods.

4. *Mordellidæ.*—Antennæ short and pectinated; thorax
semicircular; body elevated and arched ; head inserted
very low ; hind legs broad and compressed ; colour
variegated ; size small.   Locality, flowers, particularly
whitethorn.

5. *Cantharidæ.*—Antennæ generally filiform ; mandibles
ending in a simple point ; palpi filiform ; head dilated
behind the eyes, and suddenly narrowed into a short
neck ; elytra bent down at the sides; colour variegated; size moderate.   Locality, vegetable substances.

6. *Salpingidæ.*—Antennæ inserted at the base of the eyes;
palpi short and filiform ; legs slender ; head produced
into a snout ; body oval or oblong and depressed ;
colour bright ; size small.   Localities, bark of trees,
and flowers.

7. *Ædmeridæ.*—Antennæ moderately long, and filiform ;
mandibles triangular, terminating in a bifid point ;
head more or less elongated ; body often long and
narrow ; colour generally lively; size moderate.   Locality, flowers.

8. *Melandryidæ.*—Antennæ short and filiform ; mandibles

short ; maxillary palpi three-jointed, end joints very large, and often serrated ; body long and depressed ; hind legs formed for leaping in some species ; mouth not prolonged ; colour obscure ; size small. Locality, bark of trees.

9. *Cistelidæ.*—Antennæ not concealed by the margin of the head ; mandibles sometimes entire, at others cleft at the tips ; palpi filiform, or ending in a hatchet-shaped joint ; elytra somewhat soft ; size small. Localities, flowers and hedges.

10. *Diaperidæ.*—Antennæ short, and perfoliate or moniliform, suddenly thickened at the tips ; legs short ; body square or rounded ; colour variegated ; size small. Locality, on fungi.

11. *Tenebrionidæ.*—Antennæ short, moniliform, or slightly thickened at the tips ; mandibles short and triangular, the tip bifid ; palpi, last joint hatchet-shaped ; wings fit for flight ; body generally oblong ; colour obscure ; size small. Localities, on the ground or under stones.

12. *Blapsidæ.*—Antennæ rather short and filiform ; mandibles bifid ; maxillæ armed with a claw ; wings wanting ; legs slender ; colour black ; size moderate. Locality, damp places.

SECTION III.

TETRAMERA.—With apparently but four joints in each tarsus, the fifth being very minute.

1. *Bruchidæ*.—Antennæ filiform or slightly thickened at the tip, serrated or pectinated ; palpi filiform ; elytra shorter than the body ; hind legs often very large ; rostrum or snout short and broad ; body oval ; colour blackish or varied ; size small.　Locality, flowers.

2. *Attelibidæ*.—Antennæ straight; palpi minute and conical ; rostrum cylindrical and curved ; body oval, narrow in front ; colour generally black, varied with red or yellow spots ; size small.　Locality, various plants.

3. *Curculionidæ*.—Antennæ elbowed, and terminated in a club ; palpi small and conical ; rostrum or snout very long; colour generally very splendid; wings frequently wanting ; size very small.　Localities, vegetables and seeds.

4. *Scolytidæ*.—Antennæ, basal joint long, the end ones forming a solid mass ; palpi minute and conical ; maxillæ thin, broad and spined ; rostrum or snout short ; body oblong and convex ; colour obscure ; size small.　Locality, trees.

5. *Prionidæ.*—Antennæ long, filiform or setaceous ; mandibles very large ; palpi long ; upper lip obsolete ; thorax sometimes toothed ; head not narrowed into a neck ;
colour dark ; size small. Locality, trunks of trees.

6. *Cerambycidæ.*—Antennæ very long, and never serrated ;
upper lip distinct ; mandibles robust ; body mostly
long and narrow ; colour beautifully varied ; size large.
Localities, forests and hedges.

7. *Lepturidæ.*—Antennæ moderate ; mandibles acute at
the tips ; head inclined downwards, and elongated into
a neck ; elytra narrowed at the tips ; colour often
dark, with yellow markings ; size moderate. Locality,
trunks of trees.

8. *Crioceridæ.*—Antennæ filiform and short ; mandibles
short ; maxillæ broad ; head sunk in the thorax ; body
oblong, and wider than the thorax ; colour various ;
size small. Locality, plants.

9. *Cassidæ.*—Antennæ short and filiform, inserted on the
upper side of the head ; parts of the mouth small, and
on the under side of the head ; body flat beneath,
oval, with the thorax and elytra dilated at the sides
into a broad, flat margin ; head nearly concealed ;
colour diversified ; size small. Locality, plants.

10. *Galerucidæ.*—Antennæ long, of an equal thickness, and inserted close together ; palpi small and conical ; body oval or rounded ; colour obscure in general, though some have a metallic lustre ; size small. Locality, leaves.

11. *Chrysomelidæ.*—Antennæ widely apart, short and slightly thickened at the tip ; mandibles notched ; body hemispherical ; palpi short ; legs rather thick ; colour highly metallic ; size small. Localities, plants and vegetables.

SECTION IV.

Trimera.—Tarsi apparently composed of three joints.

1. *Coccinellidæ.*—Antennæ often very short, with a flattened club ; mandibles bifid at the tips ; maxillary palpi, terminal joint hatchet-shaped ; thorax very short ; body round and convex ; colour red or yellow, with black spots, or black with white, red, or yellow spots ; size small. Locality, gardens.

Order II. STREPSIPTERA, *Kirby.*

Characters.—Wings, posterior pair very large, folding longitudinally, anterior pair transformed into short slender appendages; *mouth* with slender acute jaws, and two palpi; *tarsi* two- three- or four-jointed; *antennæ* two; *body* long and narrow; *eyes* very large; *thorax* very large; *legs* moderate; *size* small.

This Order has but one family—*Stylopidæ*,—sufficiently described in the characters of the order.

ORDER III. DERMAPTERA, *Leach.*

Characters.—*Wings* two, with numerous transverse and longitudinal folds; *elytra* leathery, only partially covering the wings, and uniting with a straight suture or opening; *mouth* with transversely movable jaws; *body* forked at the end, long, narrow, and flattened; *head* of moderate size; *eyes* small; *clypeus* distinct; *palpi* filiform.

This Order has only one family—*Forficulidæ.*

ORDER IV. DICTYOPTERA, *Leach.*

Characters.—*Wings* two, longitudinally folded; *elytra* coriaceous, and one folding over the margin of the other obliquely; *antennæ* long and setaceous, with from fifty to a hundred and fifty joints; *mandibles* short, but strong, horny, and toothed at the tip;

maxillary palpi long, the last joint hatchet-shaped; *legs* long; *tarsi* five-jointed.

This Order has but one family—*Blattidæ*. It is frequently comprised in *Orthoptera*.

ORDER V. ORTHOPTERA, *Olivier*.

Characters.—*Wings* two, folded longitudinally; *elytra* coriaceous, one folding over the margin of the other longitudinally; *mouth* with transversely movable jaws.

1. *Achetidæ*.—Wings and wing-covers held horizontally when at rest; wings of large size, forming long filaments extending beyond the body; antennæ often longer than the body; body robust; eyes large; body terminated by setæ (bristles); tarsi slender; colour generally brown. Locality various, according to the species.

2. *Gryllidæ*.—Wings and wing-covers disposed like a sloping roof, or deflexed when at rest; antennæ very long and slender; wings very large and delicate; tarsi broad and fleshy; size large; colour generally green. Locality, plants.

3. *Locustidæ*.—Wings and wing-covers deflexed; antennæ

short ; body robust ; mandibles strong and toothed ; size small.

ORDER VI. NEUROPTERA, *Linnæus.*

Characters.—*Wings* four, generally large, and of equal size, membranaceous ; *mouth* with transversely movable jaws; *body* not armed with a sting.

1. *Psocidæ.*—Posterior wings small ; antennæ slender, long and setaceous; labial palpi almost obsolete; mandibles horny and toothed ; maxillæ elongated; body short and ovate ; wings deflexed, with conspicuous veins, often coloured ; legs long and slender ; tarsi two- or three-jointed ; size minute. Localities, trunks of trees, palings, old walls, books, &c.

2. *Perlidæ.*—Posterior wings large ; antennæ nearly as long as the body ; labial palpi developed ; mandibles rudimental ; body oblong and depressed ; head flat ; tarsi three-jointed ; size moderate.

3. *Ephemeridæ.*—Wings, hinder pair very small; antennæ small ; mouth almost obsolete ; body long and slender, with elongated setæ at the extremity ; head small ; clypeus sometimes large, and closing over the mouth ; legs slender ; tarsi five-jointed ; size small. Locality, near water.

4. *Libellulidæ.*—Wings large and equal ; antennæ short
and slender ; mandibles horny, thick, and strong ;
maxillæ long and toothed ; body very long and slender ; eyes very large, uniting at the top of the head ;
legs short, slender, and armed with spines; tarsi three-
jointed ; colour various and beautiful ; size large.
Locality, standing water.

5. *Hemerobiidæ.*—Wings broader than the last ; antennæ
long and filiform ; jaws horny and acute; maxillæ long;
body shorter than the preceding family ; head small ;
eyes prominent, and often gold-coloured; legs slender ;
tarsi five-jointed ; colour bright, often green ; wings
with a prismatic lustre ; size small.

6. *Sialidæ.*—Wings, anterior pair large, hinder pair
smaller ; antennæ long and filiform ; jaws horny and
toothed ; prothorax large and square ; tarsi five-
jointed ; colour in some species dull brown ; size
moderate. Locality, near water.

7. *Panorpidæ.*—Wings equal ; head prolonged into a rostrum ; antennæ long and slender ; mandibles small
and toothed ; maxillæ two-lobed ; clypeus pointed at
the tip ; body long and slender ; prothorax forming a
collar; legs long and slender; tarsi five-jointed; body,

the males have the sixth and seventh segments thin, and curved, the eighth thickened and armed with forceps, the female has minute filaments at the tip. Locality, hedges in damp situations.

8. *Raphidiidæ.*—Wings equal and of moderate size, deflexed when at rest; antennæ slender and filiform; head oval and flat; mandibles powerful; maxillæ with two lobes; prothorax much elongated, forming a neck; legs slender; size small   Locality, woods.

ORDER VII. TRICHOPTERA, *Kirby.*

Characters.—*Wings* four, membranous, the anterior pair hairy, the hinder pair the largest; *mouth* unfitted for mastication; *mandibles* rudimental; *prothorax* very short; *antennæ* often much longer than the body, slender and setaceous; *palpi* long and slender; *colour* brown or grey; *size* small or moderate.

This Order has one family—*Phryganidæ.*

ORDER VIII. HYMENOPTERA, *Linnæus.*

Characters.—*Wings* four, membranous, the posterior pair smaller than the anterior; *mouth* with horny jaws, lower lip sheathed by the maxillæ, which form a rostrum; *body* armed with a sting; *tarsi* generally five-jointed; *legs* generally long and slender.

1. *Tenthredinidæ.*—Wings large; antennæ variable in form, but generally short; mandibles long, horny, and toothed; maxillæ long and membranous; thorax generally broader than the head; legs of moderate length; body of female furnished with a pair of saws; colour dark, varied with white, red, or yellow; size moderate. Locality, various plants.

2. *Uroceridæ.*—Wings membranous; antennæ filiform or setaceous; mandibles short, but very strong; maxillæ elongated; thorax elongated; body long and cylindrical; female furnished with a borer; colour various; size moderate. Locality, fir-woods.

3. *Cynipidæ.*—Wings with few nervures; antennæ inserted in the middle of the face, of moderate length and slender; mandibles short and robust, toothed at the extremity; maxillæ elongated; thorax oval and thick; body much compressed; ovipositor long and spiral; size small.

4. *Evanidæ.*—Wings veined; antennæ filiform or setaceous; mandibles toothed; hind legs longest; body attached by a peduncle (stalk); ovipositor straight; size small.

5. *Ichneumonidæ.*—Wings veined; antennæ generally slender and filiform; mandible small and curved; thorax oval; legs long; body long and cylindrical;

ovipositor straight; colour black, varied with red, yellow, or white; size small.

6. *Chalcididæ.*—Wings nearly destitute of veins; antennæ
   elbowed, and generally thickened at the tips; mandibles
   broad and horny, toothed at the top; thorax oval;
   ovipositor sometimes concealed, at others long; body
   slender; colour metallic; size very minute.

.7. *Proctotrupidæ.*—Wings almost destitute of veins; antennæ elbowed, and variable in length; mandibles
   sickle-shaped; maxillary palpi long and pendulous;
   legs long; thorax oblong; body ovate; ovipositor
   sometimes long and curved, sometimes concealed;
   colour black; size minnte. Locality, grass.

8. *Chrysididæ.*—Wings, veins indistinct; antennæ filiform
   and elbowed; mandibles long and curved; thorax
   oblong; body oblong, shining, and attached to the
   thorax by a peduncle; ovipositor composed of several
   segments of the body terminated with a minute sting;
   colour very brilliant, blue and green in the head and
   thorax, ruby or copper-coloured in the body; size
   moderate or small. Localities, flowers, sand-banks, etc.

9. *Crabronidæ.*—Wings generally developed in both sexes,
   anterior pair not folded; head large and nearly square;

mandibles slightly curved ; antennæ straight ; legs
fitted for walking and burrowing, not for collecting
pollen ; body of various forms ; appearance very like
Wasps ; size moderate.   Localities, posts, palings, etc.

10. *Sphegidæ.*—Antennæ long and filiform ; mandibles
long, curved, and acute at the tips ; legs formed for
walking and burrowing, and very long ; body long,
often attached by a peduncle ; sting powerful ; colour
metallic.   Locality, sand-banks.

11. *Formicidæ.*—Wings wanting in the neuters ; antennæ
long and slender in the males, shorter and thicker in
the females, basal joint very long, strongly elbowed ;
mandibles horny ; body small; head triangular; thorax
in the males and females has no contraction in the
middle, which is the case with the neuters ; sting in·
some species ; body sometimes joined by a peduncle ;
colour various ; size small.

12. *Eumonidæ.*—Wings in both sexes, and folded in their
entire length ; antennæ often recurved and hooked ;
thorax short and truncate ; body joined by a peduncle,
and having a sting ; not clothed with hairs ; legs
without spines ; colour generally black and yellow ;
habits solitary.

13. *Vespidæ.*—Wings folded longitudinally; antennæ not hooked, as in the last family; mandibles truncate and toothed; maxillæ terminated by a small projection; body rarely joined with a peduncle, and having a sting; colour black and yellow; habits social.

14. *Andrenidæ.*—Antennæ elbowed; mandibles simple, or terminated by two notches; hind legs clothed with hairs to collect pollen; parts of the mouth short; habits solitary.

15. *Apidæ.*—Antennæ often elbowed, basal joint long; parts of the mouth long like a proboscis, and folded downwards in repose; body armed with a sting; habits solitary or social, according to the genera.

ORDER IX. LEPIDOPTERA, *Linnæus.*

Characters.—*Wings* four, covered with minute scales; *mouth* consisting of a long, spirally coiled organ of two pieces, representing maxillæ, with small jointed appendages at the base; *eyes* large; *antennæ* long, and variable in form; *thorax* ovate; *tarsi* five-jointed.

SECTION I. DIURNA.

1. *Papilionidæ.*—Antennæ with a distinct club; wings

broad, in some species lengthened behind into two tails ; palpi very short ; legs all formed for walking ; claws distinct ; size various. Some genera of this family are distinguished by having a groove in the hinder wings for the reception of the body, in others this is wanting; colour orange, brimstone, and white, the latter predominating, or black and yellow.

2. *Nymphalidæ.*—Antennæ with distinct club ; wings robust and grooved to receive the body ; palpi rather long ; fore legs rudimental ; claws entire ; colours beautifully varied, many of the species spotted with silver ; size moderate.

3. *Lycænidæ.*—Fore legs fitted for walking ; claws very minute and entire ; wings small and weak ; colours various according to the genera, blue and copper prevailing.

4. *Hesperidæ.*—Legs of uniform size in both sexes ; hind legs spurred ; lower wings horizontal when at rest ; antennæ wide apart at the base, in some species terminated with a hook ; maxillæ very long ; colour principally brown ; size small.

SECTION II. CREPUSC LARIA.

1. *Sphingidæ.*—Wings small ; antennæ prismatic and terminated by a little feather ; spiral tongue very long ;
body long, and acute behind ; mandibles minute ;
colour very varied ; size large.
2. *Anthroceridæ.*—Wings bent down in repose ; legs long,
hinder pair with spurs ; antennæ simple and fusiform,
or thickened near the middle, nearly setaceous and
pectinated in the males ; maxillæ very long.
3. *Ægeridæ.*—Wings more or less transparent ; antennæ
simple, fusiform, or thickened at the tips, which are
generally terminated by a small tuft of hairs ; body
elongated and terminated by a brush.

SECTION III. NOCTURNA.

1. *Hepialidæ.*—Wings long, deflexed in repose ; antennæ
short and filiform ; body long ; spiral tongue very
short or wanting ; thorax never crested.
2. *Bombycidæ.*—Wings large, either extended horizontally
or deflexed at the sides ; antennæ often strongly
pectinated in the males ; thorax not crested ; mouth
almost obsolete ; body thick and hairy ; prevailing
colour grey or fawn-colour ; size large.

3. *Arctiidæ.*—Wings deflexed ; antennæ strongly serrated ; spiral tongue very small.
4. *Noctuidæ.*—Wings moderate in size and often marked
   with peculiar ear-shaped spots ; antennæ almost always simple ; thorax stout and often crested ; spiral
   tongue long ; body elongated and clothed with scales ;
   colour generally sombre.
5. *Geometridæ.*—Wings large, held horizontally ; antennæ
   variable ; parts of the mouth short ; body slender ;
   thorax not crested ; legs slender ; colour more beautiful than the last, many species having a broad waving
   band across the fore wings ; size moderate.
6. *Pyralidæ.*—Wings moderate in size, forming a triangle
   when in repose ; antennæ simple ; mouth small ;
   labial palpi often long ; legs very long, and in some
   species ornamented with bunches of hair ; body slender
   and elongated ; size small.
7. *Tortricidæ.*—Fore wings broad and entire, forming a
   triangle when at rest ; antennæ generally simple ;
   spiral tongue short ; labial palpi forming a short beak
   in front of the head ; legs spurred ; body slender ;
   thorax rarely crested ; size minute.
8. *Yponomeutidæ.*—Wings often long and narrow ; antennæ

long, slender and simple ; palpi long and slender ; legs moderately long and spurred ; body long and slender ; colour often white, or slate with black spots ; size minute.

9. *Tineidæ.*—Wings narrow, much folded in repose ; antennæ of moderate length, slender and simple ; spiral tongue short ; legs spurred ; thorax rarely crested ; body long and slender; colour white, buff, or yellow, marked with lines ; size small.

10. *Aleucitidæ.*—Wings cleft into feathery rays, and carried horizontally ; antennæ long, slender, and setaceous ; spiral tongue long ; legs long and slender ; size small

ORDER X. HOMOPTERA, *M'Leay.*

Characters.—*Wings* four, membranous and deflexed ; anterior pair the larger, and not wrapping over in repose ; *body* convex ; *antennæ* mostly very short and terminated by a bristle ; *mouth* arising from the hinder part of the head ; *mandibles* and *maxillæ* enclosed in a canal.

SECTION I.

TRIMERA.—Tarsi three-jointed ; antennæ minute.

1. *Cicadidæ.*—Wings, anterior pair long and narrow ; an-

tennæ inserted between the eyes, of seven joints, gradually thickening to the tip; rostrum elongated; head short; eyes large; legs not formed for leaping; body provided with two internal plates, which enable the males to make the monotonous music for which they are famed.

SECTION II.

DIMERA.—Tarsi two-jointed; antennæ moderate in length and filiform.

1. *Psyllidæ.*—Wings membranaceous and deflexed, the anterior pair being of firmer consistence than the hinder; antennæ long and filiform, terminated by setæ; rostrum short, placed between the fore legs; thorax very large.

2. *Aphidæ.*—Wings much deflexed, fore wings much larger than the hinder pair; antennæ sometimes of great length; rostrum varying in length, sometimes half as long as the body; thorax oval; body short and convex.

SECTION III.

MONOMERA.—Tarsi one-jointed; antennæ six- to twenty-five-jointed.

1. *Coccidæ.*—Fore wings large, the hinder pair very minute,

like halteres ; antennæ much longer in the males than in the females ; the latter have no wings ; body elongated in the males, short and globose in the other sex ; size very small.

Order XI. HETEROPTERA, *Latreille.*

Characters.—*Wings* four, anterior pair larger than the posterior, wrapping partly over each other; basal part coriaceous, the rest membranaceous; *antennæ* generally long and filiform; *mouth* arising from the front of the head.

SECTION I.

Hydrocorisa.—Those residing in water.

1. *Notonectidæ.*—Hind legs long and hairy, forming a pair of oars ; fore legs short and curved, for catching their prey ; elytra deflexed ; antennæ inserted beneath the eyes, and four-jointed ; rostrum short and thick, capable of inflicting a painful wound ; body boat-shaped.

2. *Nepidæ.*—Fore legs raptorial, formed for seizing their prey, the others made for creeping ; antennæ short and variable ; rostrum short and robust ; body depressed and sometimes furnished with two long filaments ; head small ; colour dull brown in both families.

SECTION II.

Aurocorisa.—Those breathing the air.

1. *Acanthidæ.*—Legs long and slender ; antennæ half the length of the body ; eyes large ; rostrum long and slender ; body oval and depressed ; tarsi three-jointed. Locality, the sea-shore and banks of rivers.

2. *Hydrometridæ.*—Legs variable in different genera, but not raptorial ; tarsi two-jointed ; antennæ long, slender, four-jointed ; rostrum of moderate length ; head as broad as the thorax ; body long and narrow. Locality, the surface of running or standing water.

3. *Cimicidæ.*—Legs moderately long and slender ; tarsi three-jointed ; antennæ four-jointed ; rostrum long ; wings reduced to a pair of short scale-like pieces ; body more or less round.

4. *Tingidæ.*—Fore legs sometimes raptorial ; tarsi often only two-jointed ; antennæ filiform ; rostrum very short ; body broad and depressed.

5. *Capsidæ.*—Legs long and slender ; tarsi three-jointed ; antennæ long, second joint often thick at the tip, and the terminal ones very slender ; rostrum long and four-jointed ; the crustaceous part of the elytra terminated

by triangular piece ; body convex. Locality, plants and trees.

6. *Lygæidæ.*—Tarsi three-jointed ; antennæ four-jointed, the terminal joint not thinner than the preceding ; rostrum moderate ; body narrow. Locality, roots of plants.

7. *Scutelleridæ.*—Tarsi three-jointed ; antennæ elongated ; rostrum long ; scutellum or shield large ; size large or moderate ; colour varied. Locality, trees and plants.

8. *Coreidæ.*—Legs long ; tarsi three-jointed ; antennæ four-jointed, the end joint thickened or elongated ; rostrum long; colour diversified; size large. Locality, trees and plants.

Order XII. APHANIPTERA, *Kirby.*

Characters.—*Wings* four, rudimental, resembling minute scaly plates applied to the sides of the body; *mouth* formed for suction; *mandibles* and *tongue* long; *maxillæ* short; *antennæ* minute; *body* compressed.

1. *Pulicidæ.*—Body covered with a hard shining skin, clothed with sharp bristles on the back and legs ; no marked separation between the three parts of the body ; head small ; eyes small and round ; legs long, formed for leaping.

ORDER XIII. DIPTERA, *Aristotle.*

Characters.—Wings two, membranous, not capable of being folded; *halteres* two, small filaments clubbed at the tip; *mouth* with a fleshy proboscis forming a canal, enclosing several lancet-like organs; *tarsi* five-jointed; *thorax* short and robust; *head* distinct and attached to the thorax by a very short and narrow neck; *antennæ* very variable, inserted in the forehead.

1. *Culicidæ.*—Head small; antennæ slender, filiform, as long as the thorax, plumose in the male; rostrum nearly half the length of the insect, and slightly thickened at the tip; thorax oblong; body long and slender, on which the wings lie when at rest; legs long and slender.

2. *Tipulidæ.*—Head small; antennæ variable; proboscis very short; palpi longer than the proboscis; body and legs long and slender.

3. *Stratiomidæ.*—Antennæ various, of not more than six or seven joints, in some species the joints appear fewer in number, and are terminated by a seta or bristle; mouth mostly rudimental; wings placed on the body when at rest; body broad and depressed; colour prettily variegated. Locality, flowers in damp situations.

4. *Tabanidœ.*—Trophi greatly developed ; antennæ, third joint very large, remaining joints closely united and slender; eyes very large, and often beautifully coloured; wings extended horizontally at the side of the body ; body triangular and depressed ; thorax thick; size large. Locality, woods and pastures.

5. *Bombyliidœ.*—Proboscis long ; antennæ close at the base, third joint fusiform and flattened ; thorax much elevated, making the head appear low ; wings apart when at rest ; body often thickly covered with hairs, short and thick. Locality, flowers.

6. *Empidœ.*—Proboscis long and folded beneath the breast; antennæ as long as the head, third joint longest ; head small and round ; eyes large ; wings large ; body narrow; legs of moderate length ; size small. Locality, flowers.

7. *Hybotidœ.*—Proboscis short; head small and spherical; eyes occupying nearly the whole surface ; thorax very elevated; body narrow ; size small. Locality, flowers.

8. *Asilidœ.*—Proboscis as long as the head, which is depressed ; antennæ, third joint long, terminated with a seta ; thorax narrow in front ; body long, clothed with stiff bristles ; size large.

9. *Therevidæ.*—Proboscis ending in two large lobes; antennæ, third joint ovate, terminated by a seta; body conical, and clothed with silky down; size moderate. Locality, trees.

10. *Dolichopidæ.*—Tongue short and acute; antennæ short, ending in a small oval joint with a long seta; wings lying on the body when at rest; legs long; body compressed and curved at the tip; colour brilliant and metallic; size small. Localities, walls, trunks of trees, or near water.

11. *Syrphidæ.*—Proboscis long, membranous and elbowed; antennæ three-jointed, the third being the largest, ending in a seta; head round, almost covered by the eyes, and sometimes produced into a kind of beak in front; hind legs often thickened and toothed; body never curved at the end; colour varied; size large or moderate.

12. *Conopidæ.*—Proboscis long and elbowed, of various shapes, cylindrical, setaceous, etc.; antennæ with very short setæ; body curved at the end; colour prettily varied. Locality, plants and flowers.

13. *Muscidæ.*—Proboscis short, thick, membranaceous and distinct; antennæ with the third joint the longest;

legs and wings moderately long ; body generally short and robust, and not curved ; size small.

14. *Œstridæ.*—Mouth obsolete ; antennæ very short, last joint rounded ; body hairy.

15. *Hippoboscidæ.*—Head distinct and circular, closely united to the thorax ; eyes often very large ; clypeus or anterior part of the head distinct, with the antennæ immersed in the anterior angles ; mouth forming a kind of rostrum ; body short, depressed, and clothed with bristles ; wings with halteres, but sometimes wanting ; size small.

The preceding families, from which I have omitted those containing only exotic or rare genera, are again divided into genera and species ; but the utter impossibility of entering into an explanation of these divisions in an elementary work of this nature, will be evident from the fact, that the subdivisions of the order *Coleoptera* would alone occupy a space equal to half the present volume. When sufficient information has been acquired of the orders and families, and a general idea obtained of the study, those works may be read to advantage which enter fully into the generic and specific distinctions : at present they would only confuse the young student.

E

The description of the family distinctions in this chapter are principally taken (though much simplified and abridged) from the valuable works of Westwood, which I would strongly recommend, together with those of Kirby and Spence, to the notice of students who wish for more extended information ; the former treating with great clearness on the difficult subject of classification ; the latter, particularly that known as the "Introduction to Entomology," detailing, in the most lively and interesting manner, the natural history, beauty, and peculiarities of this portion of animated nature. To the latter work I am deeply indebted for much useful and entertaining information, and the frequent allusions to it in this little volume will, I hope, lead my readers to a perusal of its pages, affording them delight and profit by its varied and circumstantial history of Insect Life.

CHAPTER III.

MARCH.

THE insects which make their appearance this month are principally those, personally well known to even the least observing of Nature's students; yet we shall probably find that many interesting particulars may be learned, even of the Bee and the Ant; and should any of my readers think they "know all about such common insects," they will be surprised to hear that the most eminent naturalists have spent years in investigating the habits of these active and industrious little creatures, and yet confess themselves still ignorant of many particulars concerning them. Before proceeding to the description of species, I shall give a short account of the families or groups to which they belong; thus affording a little general information, where specific detail is necessarily imperfect, from the immense number of our native insects.   As the Bee is the most useful to man, and the group, in its extended signification, requires some explanation, it shall be placed at the head of the list.

Bee is the English name for all those hymenopterous insects which compose the Linnæan genus *Apis*,—now formed into two groups, namely, the *Apidæ* and *Andrenidæ*, distinguished from each other by the parts of the mouth, which in the *Apidæ* are long, forming a trunk, folded under the head and breast when in a state of repose, and stretched forward when the insect is robbing flowers of their nectar; the mouth of the *Andrenidæ*, or rather the parts forming that organ, are short.   Another difference consists in the latter having only males and females in their societies, whilst the former possess the neuters, or working Bees, in addition.

The family *Apidæ* is that which will now engage our attention ; it is divided into two groups, the Solitary and the Social Bees.   In the former, the tibia, or shank of the hind leg, in the female, is clothed with hairs ; whilst in the latter, with the exception of the queen, this portion of the leg is broad and concave, forming a trough to carry home pollen for the use of the community : this may be noticed as one of the numerous instances of design which call for our admiration when studying the works of God. The Solitary Bees form four sub-families, divided into many genera, to enter into a description of which, would be foreign to the design of this sketch; it is sufficient to name some of those whose designations are more familiar ; such as the Upholsterer and Mason Bees, the Leaf-cutter and Cuckoo Bees, all belonging to the group which, from their habits, are called *solitary.* The Social Bees consist of the genus *Bombus,* Humble-bee, and *Apis,* Hive-bee, with some exotic genera. The first of these make their nests in various situations ; some digging to a considerable distance underground, as *Bombus terrestris ;* others selecting a crevice in a heap of stones, as *Bombus lapidarius ;* whilst the *Bombus muscorum* places its nest on the surface of the ground, covering it with moss or dried leaves. These nests are constructed by one solitary female, who forms in them a number of cells, in which she deposits her eggs. It is a curious fact that the eggs first hatched are all neuters, or working Bees, which are thus ready to assist their parent in the cares of the infant state : the eggs hatched at a later period are male and female. When the cocoons, in which the grubs undergo their transformation, are empty, they are not allowed to remain useless, but are said to be employed in holding pollen.

The tongue of the Humble Bee only differs from that of the Hive Bee in being furnished near the tip with a greater number of long hairs, forming a brush, and enabling the insect to collect a much larger quantity of honey at a time. Towards August, from three to eight females are produced, which are much larger than the others, and are destined to survive the winter, all the rest of the community dying at the approach of cold weather. The former alone possess the instinct to conceal themselves in crevices of trees and walls, and on the arrival of spring they construct a nest, and perform all those labours, for which they were preserved from the destruction of their species. The other genus, *Apis*, is now restricted to the domestic Bee, *Apis mellifica*, which will be described in its proper place.

The *Aphidæ*, or *Plant Lice*, a family only too familiar to those who delight in their garden, begin to appear during this month, and, when in great numbers, considerably diminish the beauty of many plants. The insects of this family live entirely on vegetable matter, and the loftiest tree is no more exempt from their ravages than the humblest weed ; they principally attack the foliage, and are always found on the under part of the leaf, preferring this, not only on account of its being more tender, but as affording protection from the inclemencies of the weather. Occasionally the root is the object of their choice, that of the Lettuce being often observed so thickly beset with one species, that the whole crop has been rendered of little value. These insects are sometimes winged, at others destitute of these appendages; in the spring they are viviparous (that is, produce their young alive), and in the autumn oviparous, depositing eggs, like other insects, in places where they are hatched the following spring. There are six or seven generations during the summer, which accounts for their wonderful increase; and in the autumn three more are produced. If the *Aphides* had not many enemies, their increase would be so great as materially to lessen vegetation. The larvæ and perfect insect resemble each other so much as to be distinguished with difficulty, except by size; the females are generally apterous (that is, without wings); the antennæ are filiform, and seven-jointed; the tarsi two-jointed, the first very short. This family forms part of the order *Homoptera*.

The interesting and well-known family, *Formicidæ*, or Ants, merits notice more extended than would be afforded by the mere naming of the species. The members of it resemble Bees in many of these particulars which distinguish them from most other insects,—for example, in having the largest part of their community composed of workers, or imperfect females. These useful creatures seem to have the sole charge of domestic affairs ; for though the parent Ant lays the eggs, they are always under the immediate and solicitous care of the workers, who regulate the degree of heat and moisture necessary for hatching them ; and when the young grubs appear, it is on these attentive nurses that the duty of feeding them devolves, which they do, either with their own half-digested food, or some peculiar fluid secreted for that purpose. It seems probable that more than one of these kind foster-mothers is required for each larva. When the larvæ are full-grown, they enclose themselves in cocoons, in which they undergo the rest of their change, and the attachment of the workers to these cocoons is even greater than to the eggs or young larvæ. They may be seen bringing them out every fine morning, taking them in when the heat is too intense, sheltering them from rain, and, although only half the size of their nurslings, running with them as fleetly as if they had no weight whatever. Their love for these partially-developed young is so great, that they are said to invade other Ant-hills, to carry off all the cocoons they find, simply, as it seems, for the pleasure of nursing and feeding them. When the supply of food fails, Ants seek their dwellings, and in the inmost recesses they cluster together, and pass the inclement season in a state of torpor. Their food varies greatly; they are fond of sugar, gum, and all the sweet exudations of trees, but their staple food is animal matter,—either those larvæ injurious to plants, or small animals, whose dead bodies would otherwise taint the air : thus, like every other living creature whose habits are known, they are found to be beneficial to mankind, and to the animal creation in general. Many birds use them almost exclusively as food for their young; pheasants and partridges seem particularly fond both of Ants and their cocoons. They are winged at one season of the year only, and the female soon rids herself of these appendages, appearing then inspired solely with the ambition of founding a nation ; while labourers from large Ant-hills are continually on the watch for these wandering mothers, whom they bring home to their extensive cities, or raise new buildings for the reception of herself and offspring.

Ants belong to the order *Hymenoptera*, like the Bee and Wasp, from which they, however, differ in many essential particulars: their workers are wingless, and the females destitute of a sting ; the difference in their appearance is obvious. There are about a dozen British species, some of which will be mentioned hereafter.

As one of the Crickets may be both heard and seen about this time, the family of which it is a member, and those to which it is allied, claim our attention, as they are often confused with each other by the inexperienced. There are three families in this group of insects, named by Leach, *Achetidæ*, *Gryllidæ*, and *Locustidæ*. The first includes the Crickets, of which *Acheta domestica*, or the common House-Cricket, is an example ; the second, those Grasshoppers which have long antennæ, such as the *Gryllus viridissimus*, green Grasshopper; and the third, those with short antennæ, as the *Locusta migratoria*, the well-known destructive Locust. These families may also be distinguished by their wings: in *Achetidæ*, the wings and wing-covers are held horizontally when at rest ; in *Gryllidæ*, the wings are deflexed (bent down), the mandibles are also not so much toothed ; the *Locustidæ* have the wings also deflexed, but then they are known by their short antennæ. To this family belong a number of small species, very commonly found in grass, and to which the familiar name of Grasshopper is given.

I shall now enter a little more into detail respecting the order *Lepidoptera*, as the first species makes its appearance this month, though some, having survived the winter, may be seen on sunny days even earlier. This, the most beautiful, if not the most interesting order of insects, will, from the attractive nature of the species, occupy a considerable portion of these sketches; and it is therefore desirable to be well acquainted with its leading features. The order *Lepidoptera* includes the Butterflies, Hawk-Moths, and Moths, terms nearly corresponding to the genera *Papilio*, *Sphinx*, and *Phalæna* of Linnæus, and to the modern divisions founded on the times of flight—Diurnal, Crepuscular, and Nocturnal. One of the distinctions between these families consists in the form of the antennæ; those of the Butterfly being clavate (that is, terminated by a club); those of the Hawk-Moth, prismatic, thickest in the middle; and of the Moth, setaceous, gradually tapering to the extremity. The position of the wings, when at rest, forms another distinctive mark; those of the day-flying *Lepidoptera* being held vertically, whilst those of the Moth are never in this position, but vary considerably in their angle to the plane of position. The thorax of these latter insects is shorter and more robust than among Butterflies, and they possess a stiff bristle at the base of the under-wings, which, passing through a hook on the under side of the anterior pair, maintains them in their horizontal or inclined position. Having thus separated the Moth from the Butterfly, we must endeavour to divide the former from the Hawk-Moth. The most characteristic mark of the latter is to be found in the different form of the antennæ, which has been described above ; the wings too are longer, narrower, and of a firmer consistence than those of the true Moth, being also of smaller size, when compared to the body ; by their rapid vibration, the insect is poised in the air like a Hawk (hence its name), and as many of the species, when thus hovering, make a slight humming sound, and in this respect, as well as in the rapidity of their movements, bear some resemblance to humming-birds, they are known under the mixed name of Humming-bird Hawk-Moths.

There are also some differences observable in the Caterpillars of the three divisions. Those of Butterflies have usually sixteen legs, six of which are placed on the anterior part of the body, which represents the thorax of the future insect, and these legs are divided into segments corresponding to the parts which compose the leg of the Butterfly ; the others, called prolegs, and attached to the hinder part of the body, are soft and fleshy, seeming to be principally used for support. The caterpillars of Hawk-Moths are in the latter respect similar to those of the Butterfly ; but the prolegs of Moth larvæ vary both in number and structure, though the anterior legs are always of the same number as in the Butterfly, the most conspicuous being provided with ten prolegs, others with eight, six, four, and some of the smaller kinds with only two. The colour of the caterpillars varies so much, that no distinction can be drawn from .it : that of the Hawk-Moth is sometimes very showy, green being a prevailing hue, which is often varied with oblique stripes of yellow, blue, or purple. The larvæ of these insects are generally smooth, or covered with hard grains like seal-skin, and have a long horn near the extremity, somewhat curved and bent back. Many caterpillars of the diurnal *Lepidoptera* are also smooth, but in some cases they are covered with long hairs and spines, which latter appendages will, in some instances, pierce the skin, and in some foreign species, sting like a nettle.

The chrysalides vary equally with the larvæ and the perfect insect. Those of the Butterfly attach themselves to walls, posts, or trees, and after selecting an appropriate situation, the animal commences its curious proceeding. As they are suspended in two different ways, either perpendicularly by the tail, or horizontally by a band round the middle, the operations of the caterpillar vary accordingly. When the chrysalis is to be suspended by the tail, the first step is to cover a portion of the surface to which it is to be attached, with a layer of silken threads, so as to form a little reversed cone ; into this the animal pushes its hinder pair of prolegs, which are entangled by means of the small hooks of the foot ; the anterior part of the body is then allowed to hang with the head downwards ; soon after, it begins to bend the head up, retaining this position for some time, then permitting the head to fall again. This movement it continues for some hours, till the skin rends in the back, and the chrysalis appears : by degrees the skin dries, and is pushed towards the tail. Now the difficult task remains of extricating itself from the skin, which is its only support, and attaching itself to the silken threads, which are considerably above. In order to accomplish this, it seizes on a portion of the skin between two segments of the body, holding it as with a pair of pincers, and thus supports itself till it draws the tail from its sheath ; it then takes hold of another piece, by elongating the rings of the tail, and repeating this till the extremity touches the silk, to which it adheres by a number of little hooks with which it is provided. " These operations," says Réaumur, "are most delicate and perilous: it is wonderful that an insect which executes them but once in its life, should execute them so well ; we can only conclude that it has been instructed by a Great Master."

. When the chrysalis is to be suspended horizontally, the animal commences as before, and also prepares a silken band for encircling itself near the middle. Different methods are used for fixing this belt : some caterpillars let it hang down in a loop, and then insinuate their bodies into it when it is completed ; others bend their heads back to the point where the girdle is to be fixed, and, after fastening the threads on that side, carry them over to the other, simply by turning the head in that direction.

The chrysalides of Butterflies are generally of an angular form ; those of Moths oval or elliptical, and very rarely suspended : the colour of both is usually deep chestnut-brown, though occasionally of brighter hues. Those of Moths are either enveloped in a silken cocoon, an example of which all have seen in the Silk-worm, or buried in the ground, in a hole well lined with silk to render it soft and warm, or coated with a kind of varnish. The majority of Moths have to force their way out of the cocoon—no easy operation for creatures destitute of jaws ; sometimes the perfect insect is furnished with the means of liberation, which is then effected by an acid secretion discharged on the part from which it desires to make its exit, and which then yields to the slightest pressure.

When the caterpillar provides the means of escape for the future Moth, it does so, generally, by making a circular incision near one end, leaving a small piece as a hinge ; but in some cases a more elaborate contrivance is resorted to, such as that of the Emperor Moth, which makes at one end of the cocoon a funnel-shaped opening, like the neck of a flask, composed of a series of loose threads, converging like bristles to a blunt point, and formed of strong silk, well gummed. To prevent any enemy attacking the helpless inmate through this opening, the larva forms another funnel within, in which there is no opening, but through which, being made of very elastic threads, the moth can easily effect its egress. These elastic threads also answer another purpose,—that of compressing the body of the moth as it emerges, thus forcing a fluid into the nervures of the wings, and giving them their proper expansion ; for if the pupa is *taken* out of the cocoon, the wings remain crippled.*

We have now traced the *Lepidoptera* through the three

* Kirby and Spence, "Introduction to Entomology."

different states into which they pass, and explained the great subdivisions of the order ; the student will now be better prepared to understand those species which are presented to his notice. The order *Coleoptera*, the first mentioned, was described, with its principal family distinctions, in February.

COLEOPTERA. HYDROPHILIDÆ.

HYDROUS.

Generic Distinctions.—*Antennæ* short, and clubbed at the end; *palpi* long and slender ; *mandibles* largely toothed ; *elytra* narrowing gradually behind ; *tarsi* five-jointed.

HYDROUS PICEUS. (Plate I.) Ground-colour black, inclining to olive ; margins of the elytra faintly tinged with purple and green ; antennæ and mouth reddish ; wing-cases marked with eight dotted lines ; a few yellow spots on the body, and hairs of the same colour on the breast ; the legs very dark, and fringed with reddish-brown. This is the largest of our water Beetles, and is common in the south, though becoming rare in the northern counties. The larva is very destructive to small shell-fish ; it swims with great facility, and can seize its prey without changing its position,

F

from the freedom with which it moves its formidable jaws.
The female Beetle spins a silken bag for her eggs, in which
they float about until hatched ; the perfect insect may be
found as early as January in ponds and stagnant waters.

COLEOPTERA. SILPHIDÆ.

SILPHA.

Generic Distinctions.—*Antennæ* slightly compressed, and thickening gradually from the seventh joint to the apex; *palpi,* two pair
fixed to the maxillæ and under lip ; *mandibles* not notched; *elytra*
rounded behind; *tarsi* five-jointed; *body* formed somewhat like a
shield.

SILPHA QUADRIPUNCTATA. (Plate II.) Black and shining, with the elytra and sides of the thorax light yellow or
buff, with rounded spots of black. Length, five or six lines
(the line is the twelfth part of an inch).

There are about a dozen species of *Silpha* in Britain,
most of which are pitchy, or black in colour ; they frequent
carrion, and are of infinite service in freeing the earth from
putrid substances.

COLEOPTERA. COCCINELLIDÆ.

COCCINELLA.

Generic Distinctions.—Antennæ very short, and terminated by a compressed club; *palpi* with a very large terminal joint, which is hatchet-shaped; *body* nearly hemispherical; *thorax* angulated behind; *elytra* smooth; *tarsi* three-jointed.

COCCINELLA SEPTEM-PUNCTATA. (Plate II.) *Lady-bird.* Of a red colour, the elytra having seven black spots. This is one of the commonest species; when touched, it will emit a yellow fluid of a very disagreeable odour, which is said to be good for the toothache.

COCCINELLA VIGINTIDUO-PUNCTATA is a beautiful species, of a light yellow colour, having five black spots on the thorax, and eleven on each elytron; it is common in England. The members of this genus, of which there are about thirty species, vary from one-third to one-eighth of an inch in size, and are ornamented with various colours, black, yellow, and red being the most conspicuous. In France they are considered sacred to the Virgin, and are called " Bêtes de la Vierge," " Vaches à Dieu," etc. ; in England they are well known under the name of Lady-birds.

GRYLLOTALPA.

Generic Distinctions.—Antennæ short and slender; *wings* and *wing-covers* horizontal; *tarsi* three-jointed; *fore legs* very broad.

GRYLLOTALPA VULGARIS, or GRYLLUS GRYLLOTALPA (Plate III.), the *Mole Cricket,* is about an inch and a half or two inches long, nearly cylindrical, and of a brown colour; its legs, being formed for the purpose of burrowing in the ground, are broad and notched ; the body is terminated by two long, stiff bristles ; it inhabits the sides of canals and damp soils, in which, just below the surface, it forms long winding burrows and a chamber, neatly smoothed, in which, about the middle of May, it deposits nearly a hundred eggs. This singular insect makes a dull jarring sound, continued for a long time without interruption, like that of the Fern Owl; farmers suppose it does injury by loosening the earth round the roots of vegetables, upon which it is said to subsist, though Latreille thinks it feeds on worms and insects.

ORTHOPTERA.   ACHETIDÆ.

ACHETA.

Generic Distinctions.—Antennæ very long; *wings* and *wing-covers* horizontal ; *tarsi* three-jointed; *legs* slender.

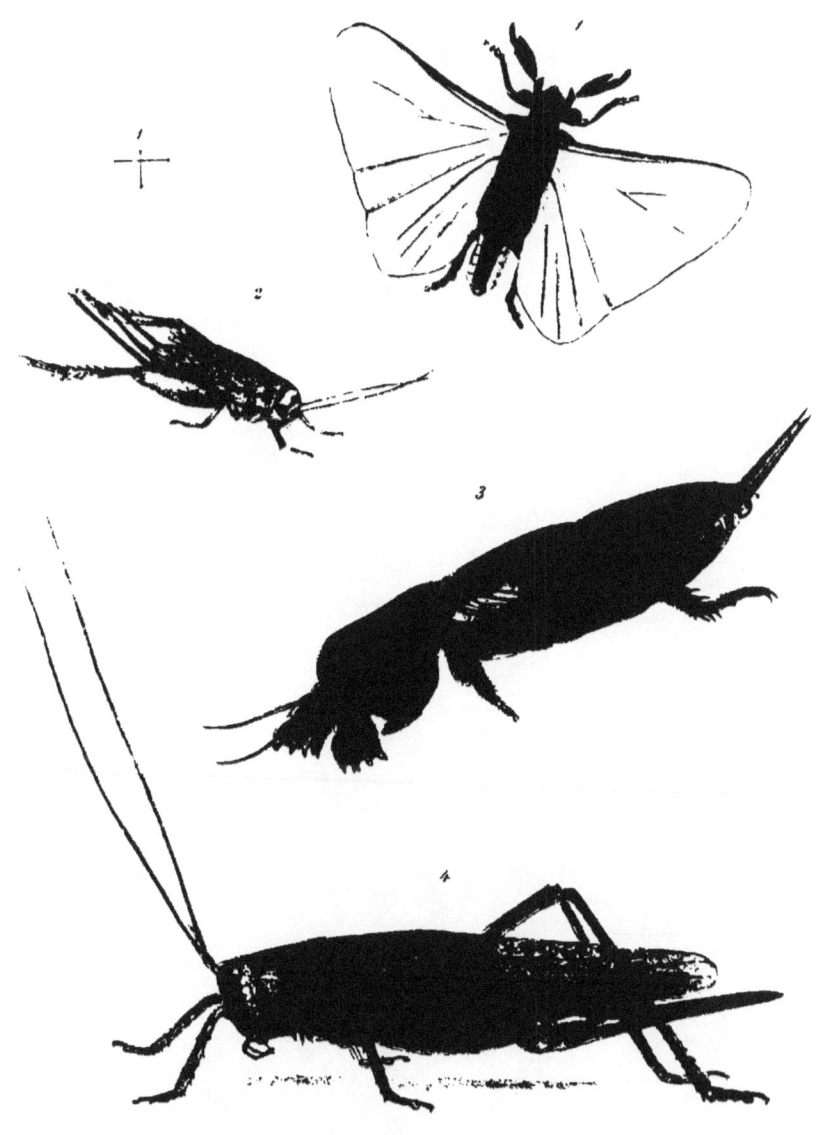

Fig..... ...   2. Acheta domestica   3. Gryllotalpa vulgaris   4. Gryllus viridissimus

ACHETA DOMESTICA (Plate III.), the *House Cricket*, needs little description ; it is of a buff colour, varied with brown ; the loud chirping noise made by the males is occasioned by the brisk attrition of their wings, the covers of which are provided with a small glassy membrane, acting like the parchment covering of a tambourine.   These insects may be found throughout the year in favourable situations.   White of Selborne says they feed on bread, yeast, salt, and any kitchen sweepings : they are fond of moisture, and will gnaw holes in wet woollen stockings or aprons hung to dry.

ACHETA CAMPESTRIS, the *Field Cricket*, is much more rare, and appears later ; the colour is black, with the base of the wing-covers yellow.   It burrows at the side of paths, making deep holes, at the mouth of which it sits, in order to seize any stray insects for food.

HETEROPTERA, NOTONECTIDÆ.

NOTONECTA.

Generic Distinctions.—*Body* subcylindrical ; *tarsi*, first joint long ; *claws* very minute.

NOTONECTA GLAUCA. (Plate XV.) *Boat Fly*. About half an inch long, of a greyish colour ; the elytra spotted with

black at the margin.   It is an aquatic insect, and derives
its generic name from the habit of swimming on its back ;
on approaching a piece of standing water, these insects may
be observed resting with the tail upwards, and the legs extended at right angles ; with the body in this position, they
are enabled to obtain a supply of air to the apertures of the
air-tubes, which are on the sides.   They produce a slight
wound with the proboscis, and feed on other insects.   Many
of my readers may have seen them by means of the gas-
microscope.

_____

HYMENOPTERA. APIDÆ.

APIS.

Generic Distinctions.—*Proboscis* long; *maxillary palpi* almost
obsolete; *body* oblong; *legs* furnished with a pollen plate, and destitute of spines; *tarsi,* basal joint oblong.

APIS MELLIFICA, the *Hive Bee,* needs no specific description, as the genus *Apis* is now restricted in England to this
species, which is one of the most perfectly social groups of
insects, and possesses the greatest share of instinct.   One
of the most striking peculiarities in the Hive Bee is the
existence of those individuals formerly regarded as neuters

but which by modern investigation are found to be females, imperfect in their organization : these constitute the great mass of the population in every hive ; to them is committed the internal economy of the society, and upon them the whole labour of the community devolves. It is also their duty to guard and protect the queen, to feed the young, and to kill the drones, or males, at the appointed season : in a single hive there are many thousands of these individuals. The perfect female, or queen, may be distinguished by her superior size and length, her brighter colour, and curved sting. Her duty principally consists in laying eggs, during which operation she is attended by a body-guard of workers, who treat her with every mark of respect. The male Bees, or drones, have no sting, the head is rounded, and the eyes larger.

As soon as the plants begin to flower, Bees are in motion for the purpose of collecting honey and wax, the former of which is a sweet limpid juice found in the nectaries of flowers, the latter is a secretion from the body of the working Bees. These different materials, being brought to the hive, are received by the labourers in waiting, who form cells of the wax, which serve as storehouses for the honey, and nests for the young. The honey is partly distributed for

present food to the inhabitants, and the remainder laid up
for winter consumption. The form of the cells is an interesting proof of instinct :—as there is no great space in a hive
which is to accommodate so many thousand insects, and as,
also, the Bees do not secrete a great quantity of wax, the
saving of both space and material is an object of considera-
tion; consequently, every Bee is endowed with instinct which
places her high on the list of geometricians, as the cells are
precisely of that form which ensures both saving of space
and material. To prove this, Réaumur, the great naturalist,
gave the following problem to König, a skilful mathema-
tician:—Among the hexagonal tubes with pyramidal bases,
to find that which can be formed with the least possible
quantity of matter. König worked the problem, and found
precisely the same angles which Réaumur had previously
ascertained to be those of the Bee's cell. It is a curious
circumstance that the design of every comb is sketched by
a single Bee, who lays the first rudiments, which are then
completed by the rest. There are three different kinds of
cells : the first are for the larvæ of the workers, and for
containing the honey ; the second, for the grubs of the
drones, which are larger than the former ; and the third, of
which there is only a small number, are destined for the

future queens : these are called royal cells, and are of quite a different form from the rest. The antennæ seem to be the organs by which the little architects regulate the shape of their wonderful buildings. The first cells are all made of the proper size and form for the working Bees ; but when the queen is going to lay male eggs, the builders immediately change the dimensions of the cell to suit the intended occupant ; they also, at certain intervals, construct royal cells, and about once in three days the queen deposits an egg, which is destined to produce a future queen.

Another extraordinary circumstance is noticed, connected with these embryo sovereigns, which is, that should the queen happen to die before having laid any royal eggs, the Bees, after much apparent consultation, select one of those formed for the working Bees ; three cells are thrown into one for its reception, the grub, when hatched, is fed with royal jelly, and a queen is produced ; whereas, had it remained in the original cell, and been fed with ordinary food, it would have turned out only a working Bee. This, and some other facts, are so contrary to what is known in other branches of Natural History, that, unless fully proved, they might be deemed fabulous.

We must now proceed to the period when the young

queens are ready to come forth in their perfect state. At this time the queen-mother appears to become infuriated ; she rushes to the cells of her royal offspring, and tearing them open, slaughters the inmates with her sting ; after killing two or three, she communicates her excitement to the workers, when a great portion of them, accompanied by their old queen, rush from the hive and seek another home. In every instance, it is the original sovereign that leads the first swarm, which, being placed in a fresh hive, again commences active labours. Meantime, the Bees remaining in the old hive take great care of the undestroyed cells, and prevent the young queens from leaving them, except at proper intervals. As soon as one of them is hatched, she proceeds to attack the royal apartments ; but the guards, who permitted the ancient sovereign to pursue her own way, do not extend the same courtesy to her successor, but surrounding the cell of her rival they bite and drive her off. Angry at this conduct, the little queen stands upright, uttering a clear shrill sound ; and no sooner is this heard than the Bees remain motionless, hanging down their heads ; again she attacks the cell, and is again driven away, till at length, giving up the contest, she quits the hive, with a second swarm. To prevent the young queens from making their appearance too soon, the workers remove a portion of wax from the cell; making it sufficiently thin to be seen through; and when the prisoner is ready to free herself, and has cut through the cocoon, they fasten up the cleft with wax, and prevent her egress; upon this she emits a distinct humming sound, which, however, excites no pity in her subjects, though they feed her with honey when she puts out her tongue for that purpose. At the proper time she is released, and leads the swarm which is then ready for her. In the autumn the males are all killed, and even those grubs that would have changed into drones participate in the general doom.

Volumes have been written, filled with interesting details respecting these useful insects; and the great difficulty, in a sketch like the present, consists in selecting from so great a mass of information that which shall interest the reader and make him acquainted with the outline of their history, within the compass to which it is limited, whilst it incites him to search for more abundant details in larger works. The "Introduction to Entomology," by Kirby and Spence, gives a very interesting account of the Bee, and from this valuable work many of the above remarks are condensed.

APHIS.

Generic Distinctions. — *Antennæ* filiform, seven-jointed ; *elytra* larger than the wings ; *body* generally horned towards the apex ; *wings* either four, or entirely wanting.

APHIS ROSÆ (Plate XV.) is generally of a green colour ; the tips of the antennæ, and horns, black ; tail pointed. These insects may be found in February, when the weather is sufficiently warm, and are produced from oval black eggs deposited in the autumn. They reach maturity in April, after twice casting their skins ; and early in June, some of the third generation, after throwing off their last covering, are found to possess four wings, which had been folded into a very small compass, but now extend in a beautiful manner to their proper form and dimensions.

APHIS HUMULI infests the Hop, and the importance of this apparently insignificant insect may be gathered from the following account given in the " British Cyclopædia of Natural History : "—" In the year 1802 the hop duty fell from £100,000 to £14,000, on account of the great increase of the *Aphis ;* in 1825, from £130,000 to £22,000 ; and in the following year, which was remarkably dry and hot, scarcely an *Aphis* was found, and the duty, which began in May at £120,000, rose to nearly £500,000; so that this little insect actually possesses a control over the British treasury amounting to hundreds of thousands."

GONEPTERYX.

Generic Distinctions.—*Antennæ* rather short and robust, thickening near the summit to an obtuse club; *palpi* compressed and projecting a little beyond the head; *wings* large, angulated, the under pair grooved to receive the body.

GONEPTERYX RHAMNI. (Plate V.) *Brimstone Butterfly*. The male is entirely bright sulphur-yellow above, and the female greenish-white; both have a small round orange spot near the middle of each wing, and some minute rust-coloured spots along the outer edge; the under side is paler than the upper, and the central marks are rust-brown and pale in the middle. The body is black above, and clothed with long silky white hairs, the under parts are yellow; antennæ reddish; the head and thorax slightly tinged with rose-colour. This insect is common in the southern counties, and also occurs in many parts of the north, but not in

Scotland. It often survives the winter, when its graceful outline and gay tints render it a pleasing herald of spring. It may often be seen vibrating like the petals of the primrose it so much resembles in colour, by the side of a sheltered wood. Though I have mentioned this pretty butterfly in the early month in which it frequently makes its appearance, yet the first broods are more usually seen in May, and the second in autumn. The caterpillar is long, naked, and of a light green colour, with many black dots on the back, and a pale line down each side; it feeds on the *Prunus spinosa*, or Blackthorn. The chrysalis is angular.

DIPTERA. CULICIDÆ.

CULEX.

Generic Distinctions. — *Antennæ* setaceous, of about fourteen joints, which form a tuft in the males; *palpi* long; *rostrum* long, and enclosing a sucker of five pieces; *wings* lying horizontally on the body; *legs* long.

CULEX PIPIENS. (Plate XV.) *The Gnat.* Body cinereous; wings transparent; antennæ of the male plumose; length of the insect three lines. The beautiful proboscis of this little creature is well worthy of notice, notwithstanding the pain and uneasiness it causes; for although it appears to the

naked eye to consist merely of a long, slender, and simple
organ, it is in fact a number of fine lancet-like pieces, enclosed in a fleshy gutter, forming an instrument admirably
adapted for suction, and provided with a poisonous liquid,
which it instils into the wound in order to make the blood
flow more freely : it must be remembered that the females
alone exhibit this thirst for blood, the males contenting
themselves (in more elegant taste) with the nectar of flowers.
The females deposit their eggs in a remarkable manner : the
larvæ are destined to inhabit water, and the perfect insect
is an aërial being ; it therefore takes its station on the extremity of a floating leaf, and, by the assistance of the hind
leg, forms the eggs into a curious boat-like mass, which
floats on the water.   These hatch in about two days, and
may be seen in the larva-state jerking about with great
rapidity ; they are small, and semi-transparent, long, and
furnished at the sides with long hairs ; the head is rounded,
having two jaws, which are kept in perpetual motion, serving to produce a current of water, which brings small
floating particles to its mouth.   They are transformed into
pupæ in about fifteen days, in which state the body is
thickened, and exhibits the rudiments of legs and wings ;
the external apparatus for breathing now consists of two

small horns, and the pupa swims with great agility by means of two small swimmerets, or little oars. There is an apparent difficulty in this little aquatic animal casting off a form with which it could not live in the air, and assuming that which would be destroyed in its future element. But Nature, or rather the God of Nature, is never at a loss for expedients; the pupa rises to the surface, when ready to enter its new condition, and elevating the thorax above the level of the water, the skin bursts, by swelling the exposed part, and the head and thorax of the Gnat are seen, while the rest of the body gradually appears. At this critical period (say the authors of the "Introduction to Entomology") the old skin acts as a lifeboat to the little animal; the observer, who sees how this little boat sinks closer and closer to the margin, feels interested for the safety of the insect, for should it upset, the tiny mariner would certainly be drowned; but the Gnat, having fixed itself perpendicularly, draws first its two anterior legs out of their case, and moving them forward, proceeds to do the same with the next pair; then resting for an instant on the surface of the water, the wings unfold themselves, are dried, and the insect flies away to enjoy its new existence.

CULEX PULICARIS. *The Midge.* Body slender; antennæ

plumose, and forked at the extremity ; wings white, about a line in length. The Midge very closely resembles the true Gnat, but differs in wanting the long proboscis. The antennæ are ornamented with whorls of hair, which, in the males, form beautiful objects. These insects do not appear till later in the year, but, being so similar to the Gnats, it is thought better to notice them together.

DIPTERA. BOMBYLIIDÆ.

BOMBYLIUS.

Generic Distinctions.—Antennæ, second joint very short, terminal one long; *palpi* very visible; *proboscis* very long; *body* short, thick, and hairy; *legs* long.

BOMBYLIUS MAJOR is about one-third of an inch long, and covered with yellowish hairs ; the wings somewhat dusky.

BOMBYLIUS MEDIUS has the wings marked with small dark spots ; the specific name bestowed on it by Linnæus is inappropriate, as it is larger than *B. major.*

BOMBYLIUS MINOR is much smaller than either of the preceding species ; the wings are quite clear.

These insects are very active, flying with great rapidity ; they hover over flowers without settling on them, but extract

G

the honey by means of their very long proboscis. Their transformations seem to be little known ; Latreille supposes them to be parasitical in the nests of other insects. The perfect fly (which very much resembles a Bee, and is hence termed the Bee-fly) may often be met with in fine weather, making a humming sound during its flight.

This family has been by some authors confounded with *Asilidæ*, from which it is however distinguished by the proboscis in the latter being very short, and the body longer. The *Asilidæ* prey on other insects, which they seize on the wing by means of their fore legs, and extract the juices with the lancet-like parts of the mouth. The larvæ in this family reside in the ground ; the body is long, the head scaly ; they are destitute of legs, but make use of the hooks on the head to draw themselves along ; they form no cocoon when about to become pupæ, but undergo the transformation in the same locality in which they have existed in the state of larvæ. The finest species is the *Asilus crabroniformis*, thus named from the resemblance it bears to the Hornet, *Vespa crabro*.

CHAPTER IV.

APRIL.

THE sunshine of this variable, though often delightful month, calls many of our insects into being. Several of the early white Butterflies may be seen flitting about, rejoicing us, as harbingers of spring, and sipping the sweets of the few flowers now in bloom ; the less delicate, but more brilliant Beetle, attracts our attention, as it runs swiftly across our path, or flies heavily during a warm evening, with its peculiar, humming sound ; the elegant Dragon Fly emerges from the watery bed in which it had passed the first stage of its existence, and hovers over the element it has so lately quitted, as though unwilling to leave its native home to try new scenes ; and though,

> " Spring is but the child
> Of churlish winter, in her froward moods
> Discovering much the temper of her sire,"

G 2

yet she sends us many of her " flying gems" as precursors of her abundant fertility, when the sun shall have tempered still more the bleak winds of March.   The beautiful family of Dragon Flies, *Libellulidæ*, is noted for the extreme elegance of the species composing it.   Kirby speaks of their dress as silky, brilliant, and trimmed with the finest lace ; and Mouffet says, " they set forth Nature's elegancy beyond the expression of art."   The French call them *Demoiselles :* our less elegant names of Dragon Fly, Horse-stinger, etc., are quite misapplied, as they are perfectly harmless.   They are generally seen skimming over standing water, seizing other insects as food ; sometimes even Butterflies become victims of these beautiful destroyers.   The eggs are laid by the female in the form of a bunch of grapes, and are soon hatched into short, thick larvæ, with six scaly legs ; on becoming pupæ, they are equally active.   In their aquatic state these insects have a curious apparatus for catching their prey, which is thus amusingly described in Kirby and Spence's admirable work :—" One of the most remarkable instruments, in which the art and skill of Divine mechanism are singularly conspicuous, may be seen in the under-lip of the Dragon Fly.   Conceive your under-lip to be horny instead of fleshy, and to be elongated, so as to wrap over

your chin, and then expand into a triangular plate attached
by a joint, so as to bend upwards again, and fold over the
face, as high as the nose, concealing not only the chin and
the first-mentioned elongation, but the mouth and part of
the cheeks ; conceive, moreover, that to the end of this
last-named plate are fixed two other convex ones, so broad
as to cover the nose and temples : that these can open at
pleasure like a pair of jaws, exposing the nose and mouth,
and that their inner edges, where they meet, are cut into
numerous sharp teeth and spines, or armed with long and
sharp claws. You will probably admit that your own visage
would present an appearance not very engaging, while concealed by such a mask ; but it would strike still more awe
into the spectators, were they to see you first open the
two upper jaw-like plates, which would project from your
temples like the blinders of a horse ; and next, having, by
means of the joint at the chin, let down the whole apparatus,
and uncovered your face, employ them in seizing any food
that presented itself, and conveying it to your mouth. This
formation so exactly resembles a mask, that if entomologists
ever went to masquerades, they could not more effectually
relieve the insipidity of such amusements, and attract the
attention of the Demoiselles, than by appearing at the

supper-table with a mask of this construction, and serving themselves by its assistance : these creatures steal upon their prey as a cat does on a mouse." They live in the water for ten or eleven months prior to their final change, and from spring until the commencement of autumn may be seen in every favourable locality ; when ready to quit the pupa state they leave the water, and, ascending the stem of some aquatic plant, allow their outward covering to become dry and brittle, when it splits down the back, and the head and legs make their appearance. The insect then seizes a twig with its fore legs, and draws out the rest of the body ; after which it remains a considerable time, until its wings acquire their full size and gauze-like appearance.

There are two families of Water Beetles seen this month, and frequently earlier,—the *Dyticidæ* and *Gyrinidæ*,—the latter remarkable for the metallic brilliancy of their covering, which distinguishes them from the *Dyticidæ*. The velocity with which they execute their evolutions on the surface of the water is truly surprising, and has obtained for them the name of "Whirligigs." One of our naturalists thus describes them :—"Water—quiet, still water—affords a place of action to a very amusing little fellow, *Gyrinus natator*, which about the middle of April we see gambolling about, on the surface of a sheltered pool; every school-boy who has angled for minnows in the brook is well acquainted with this merry swimmer in his shining black jacket; one pool commonly affords space for the amusement of several parties, yet they do not unite, but perform their cheerful circlings in separate family associations; if we interfere with their merriment, they seem greatly alarmed, disperse, and dive to the bottom, where their fears soon subside, and we see our little friends dancing as before."

Some of the beautiful Tiger Beetles may occasionally be found, their brilliant green rivalling the hue of the emerald, when seen in the sunshine; they fly swiftly, and the rapidity of their motions renders escape impossible to any insect they may attack ; they emit a fine rose-like scent. The larvæ reside in burrows of great depth, which they excavate in sand ; at the mouth of these holes they station themselves to entrap their prey, and are, like the perfect insect, furnished with hooked jaws, and six strong legs. Dr. Kidd thus amusingly describes one of the larvæ:—"Such a beauty! the *Parcæ*, sweet creatures, the *Eumenidæ*, gentle turtle-doves, were lovely in comparison ; aspect vicious, temper ferocious, jaws diabolical, stuck on the wrong way, head big, back humped, the hump adorned with two hooks."

The jaws afford one of the many proofs of design which are so frequently met with in the insect world ; they are turned upwards, contrary to the usual formation, but were they in the ordinary position, the animal, which takes its station at the mouth of its hole to catch stray insects, would be compelled to throw back the head to a great distance, in order to snap at them, whereas its present formation enables it to take them without any difficulty ; the hooks on the back are also worthy of our notice, as it is by their assistance that the insect climbs up, and retains its situation at the mouth of the cell.

The same amusing author just quoted, says of one whose hole he seems to have destroyed :—" He set to work to make another, for which purpose he used his feet and jaws, loosening the sand with his feet, and fetching it out with his jaws : in this way he got down about half an inch, and then adroitly hanging himself to the edge of the hole by his hooks, he continued his labours in this droll situation ; at last he got out of sight, and as he did not appear again, I concluded he was taking a nap after his labours."

COLEOPTERA. CARABIDÆ.

CARABUS.

Generic Distinctions.—Body elongated, often bronzed or golden green ; *head* projecting, narrower than the thorax ; *antennæ* filiform ; *thorax* less broad than the body ; *size* large.

CARABUS CLATHRATUS. About an inch long, of an oblong form ; colour dark brassy, head and thorax faintly punctured, each elytron has three elevated lines and a triple series of excavations of a copper-colour, the under side and legs black ; the wings are not adapted for flight.

CARABUS NITENS (Plate I.) is found on heaths, and scarcely yields to any exotic insect in brilliancy of lustre.

COLEOPTERA. CARABIDÆ.

ELAPHRUS.

Generic Distinctions.—Head with prominent eyes ; *antennæ* becoming thicker towards the extremity ; *thorax* nearly the same width as the head ; *tarsi*, first four joints slightly dilated in the males.

ELAPHRUS RIPARIUS. Head and thorax deeply punctured, the latter with a short groove in the middle ; elytra thickly covered with minute punctures, and four rows of purple spots, each with a ring of metallic lustre ; colour brassy-green, under side bronzed green ; legs pale yellow. This insect frequents marshy places and the margins of ponds, running very quickly.

COLEOPTERA.　STAPHYLINIDÆ.

STAPHYLINUS.

Generic Distinctions.—*Head* large ; *antennæ* short, inserted between the eyes; *body* narrow, and not covered by the elytra; *wings* very large; *palpi* filiform; *size* large.

STAPHYLINUS ERYTHROPTERUS. About half an inch long, with red elytra; found under dead leaves, or stones, or seen flying in hot sunshine.

STAPHYLINUS OLENS (Plate I.) is of a dull black colour, with pale reddish elytra; it is known under the name of the " Devil's Coach-horse." When alarmed, it opens its jaws, and throws its tail over the back; it is frequently seen running quickly across roads and sandy paths ; one very small species sometimes causes great annoyance by flying into the eyes, giving a smarting sensation from the vapour it emits; they feed on decaying vegetable and animal substances.

CHRYSOMELA.

Generic Distinctions.—Antennæ slightly thickened towards the tips; *palpi* hatchet-shaped; *body* rounded or oval; *wings* formed for flight; *colour* very brilliant.

CHRYSOMELA SANGUINOLENTA is about a third of an inch in length, of a blue-black colour, the elytra widely bordered with red.

CHRYSOMELA FASTUOSA. (Plate II.) A very pretty species of a brilliant golden-green, with a stripe on each elytron of violet-blue; it is smaller than the last, and is found on the white dead-nettle (*Lamium album*).

CICINDELA.

Generic Distinctions.—Antennæ filiform; *palpi* distinct and hairy; *wings* long and slender; *body* oblong or oval; *colour* very brilliant.

CICINDELA CAMPESTRIS. (Plate I.) A very elegant insect, of a bright grass-green above, with several white spots on each elytron.

COLEOPTERA. GYRINIDÆ.

GYRINUS.

Generic Distinctions.—*Antennæ* short and thick, forming a mass, almost like ears; *head* sunk in the thorax; *hinder legs* very short, compressed, and hairy; *fore legs* long; *body* highly polished.

GYRINUS NATATOR. (Plate I.) Nearly three lines long, of an ovate form, blue-black in colour, with a metallic lustre, the breast rather red; found in water.

COLEOPTERA. DYTICIDÆ.

DYTICUS.

Generic   Distinctions.—*Antennæ* eleven-jointed and setaceous; *head* thick, and partly sunk in the thorax; *elytra* in some species destitute of furrows in the males; *body* oval; *colour* black, tinged with olive; *size* large.

DYTICUS MARGINALIS has the body blackish-green above, yellowish-brown below, the thorax and elytra margined with yellow.   The female is furrowed; it is about an inch in length.

COLEOPTERA. PTINIDÆ.

ANOBIUM.

Generic Distinctions.—*Antennæ* of eleven joints, terminated by three larger joints; *head* short and round; *body* cylindrical and convex; *wings* strong, longer than the elytra; *colours* obscure.

ANOBIUM TESSELLATUM. Body of an obscure brown colour, with yellowish spots formed by hairs; the elytra are not striated (marked with lines).

These innocent little creatures, from a superstitious error, bear the name of Death-watch, in common with another species.

ANOBIUM STRIATUM, of a more uniform colour, and may be heard in old houses making a clicking noise in the walls and window-frames. This sound is produced by the insect raising itself on its hinder legs, and beating its head with some force against the wood on which it stands. There are eleven species in England, all of which probably possess the power of making this noise, which is merely a call or signal to its companions.

NEUROPTERA. LIBELLULIDÆ.

LIBELLULA.

Generic Distinctions.—*Head* globular; *antennæ* short; *eyes* very large; *body* long, depressed and pointed ; *wings* extended horizontally in repose.

LIBELLULA DEPRESSA (Plate IV.), *Dragon Fly*, is of a yellowish-brown colour with two yellow lines on the thorax ; the body is brown or gray, with the sides yellowish ; the wings are transparent, with a large spot of brown at the base ; length about an inch and a half ; the sexes are often of different colours, the males having the body lead-blue, whilst the females are rich yellow-brown ; in the genus *Agrion* (which some authors make a sub-family) the males are of a rich blue, with black wings, the females a fine green with colourless wings ; the body is also rounded and much more slender.

HYMENOPTERA. FORMICIDÆ.

FORMICA.

Generic Distinctions.—*Antennæ* inserted near the middle of the front part of the head ; *mandibles* strong, triangular, and dentate ; *wings* large and unequal ; *sting* wanting.

FORMICA RUFA. *Red Ant.* The neuters are blackish, with part of the head and thorax fawn-coloured ; the female has the thorax oval and fawn-coloured, and the body black ; the males and females are about four lines in length, the neuters

Plate IV.

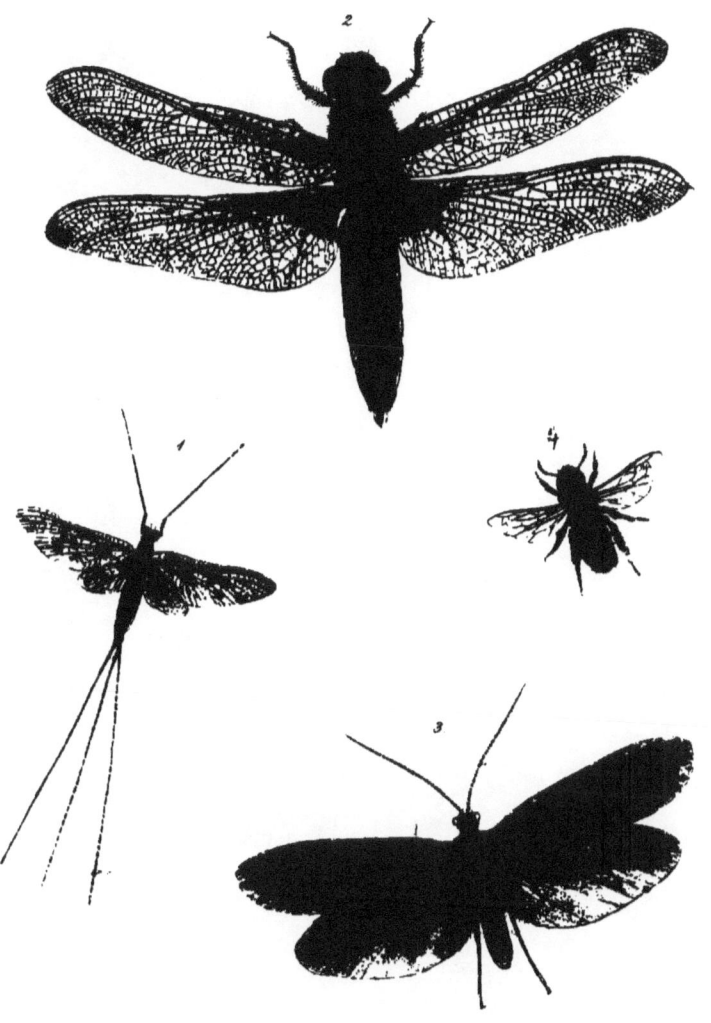

... era ... gata ... oenula depressa 3 Phryganea grandis 4 Andrena rufro-annea

rather smaller ; the wings are obscure, with yellowish nerves. The dwellings of these Ants, which are called Red, Hill, or Horse Ants, are composed of straws, wood, and earth, heaped into a cone-shape, and are found in woods ; the number of individuals of which a nest is composed is very great; when disturbed, the ants come out in immense numbers, looking very formidable, and biting severely. Their habitation is formed with much art, and differs from that of all other species, in having a great many entrances, which are all closed at sunset, when the little inhabitants shut themselves in for the night.

FORMICA NIGRA. *Black Ant.* In this species the insects are about two lines long, of a blackish-brown colour, with the mandibles and part of the antennæ paler; they dig small galleries on the sides of roads, fields, and gardens.

LEPIDOPTERA. PAPILIONIDÆ.

PONTIA.

Generic Distinctions.—*Antennæ* long, slender, and terminating in an abrupt club; *palpi* clothed with scales; *anterior wings* nearly three-cornered; *posterior wings* rounded, and not variegated beneath; *legs* alike in both sexes; *chrysalis* terminated in a beak, and attached by the tail.

Pontia Rapæ. *Small White Butterfly.* Upper surface of the wings white, slightly inclining to yellow ; the first pair have a dusky spot at the tip, extending a short way round the border ; in the male there is one, and in the female two, round spots on the disc ; the latter have also an oblong patch behind ; the hinder wings have a black mark on their anterior border. Beneath, the anterior wings have two black spots ; the tip is yellow, as are the under wings, and appears as if sprinkled with black spots ; on the anterior edge there is a streak of orange-yellow. Size about twenty lines. The caterpillar is light green, with a pale line on the back and a whitish streak on each side ; it feeds on the cabbage and turnip. There is a second flight in July.

Pontia metra. *Howard's White.* Colour entirely yellowish-white ; base of the wings blackish, and the tip of the anterior pair slightly suffused with brown ; in the male there is a single, and in the female two dusky spots, which are however sometimes wanting ; hinder wings are wholly white. Beneath, the tip of the upper wings is yellow ; the hinder wings are also yellow, sprinkled with black points. From twenty to twenty-five lines in size. There is another brood in June ; it is common in the south.

PONTIA CARDAMINES. (Plate VI.) *Orange Tip.* The primary wings in this pretty species are white, dusky at the base, with a small crescent-shaped spot in the middle, and a patch of black round the tip; the outer half of the wing is deeply tinged with orange in the male, but not in the female; the hinder wings are alike in both sexes; on the upper side they are dusky at the base, the surface presenting tracings of the markings underneath, which consist in spots of green, powdered with yellow. The caterpillar is green, with a white line on each side, and feeds on several Cruciferous plants. This insect differs slightly from other *Pontias* in the form of wings, which in the anterior pair are more rounded, and by the under side being variegated; the termininal joint of the palpi is also shorter. The *P. cardamines* measures about an inch and a half.

MELITÆA.

Generic Distinctions.—*Antennæ* with a very abrupt knob, large and flat; *palpi* long and projecting; *wings* of moderate size, anterior pair rather long and triangular; *chrysalis* suspended by the tail.

H

MELITÆ ATHALIA. *Pearl-bordered Likeness.* Colour tawny-orange, marked with several undulating black lines, running across the surface. Beneath, the anterior wings are pale-brownish yellow, with transverse streaks of black; the hinder pair with several pale angular spots near the base, edged with black; behind this there is a curved band of large pale spots, also edged with black, and near the margin two series of black crescents. The caterpillar, which feeds on plantain and heath, is black and spiny, with two rows of white dots on each segment. The insect is rare in the north, but not so in Devonshire and other southern counties.

LEPIDOPTERA. NYMPHALIDÆ.

HIPPARCHIA.

Generic Distinctions.—*Antennæ* variable in length, club generally tapering at both ends, sometimes short and abrupt; *palpi* longer than the head; *anterior wings* sometimes angular, at others rounded, and generally toothed in the hinder pair; *fore legs* very short; *chrysalis* suspended by the tail.

HIPPARCHIA ÆGERIA. *Speckled Wood.* Upper side brown, the anterior wings marked with several pale-yellow spots, having a black ocellus, or eye, with a white pupil; the hinder wings have also one or two yellow spots, and a row of ocelli round the hinder margin; the under side is pale yellow, clouded and streaked with brown; the hinder pair marked with undulating transverse lines, and a row of pale dots encircled with brown. The caterpillar is green, marked on the sides with yellow or whitish lines: it feeds on grasses. There are three flights of this insect, in April, June, and August; it may be found throughout Britain.

HIPPARCHIA MEGÆRA. *Wall Butterfly.* Anterior wings orange, inclining to brown, with the hinder margin and several transverse bands of dark brown, each wing with a large ocellus towards the tip, having a black iris and white pupil; the hinder wings are dark brown, with two transverse bands, that next the margin having a row of ocelli. Beneath, the upper wings are pale, with the brown bands faintly marked, the ocellus being larger; the under pair are ash-grey, varied with two undulating brown lines; they have also a series of ocelli and a waved band of pale yellow. The caterpillar is pubescent or downy, of a light green with a whitish line on each side. This is a common species, and occurs again in July and August.

THYMELE.

Generic Distinctions.—Antennæ terminating in an acute hook, and the club curved.

THYMELE ALVEOLUS. (Plate X.) *Grizzled Skipper.* A small species, seldom exceeding an inch in extent; ground-colour of the wings brownish black, anterior pair with straw-coloured spots; the under wings have, in addition, an irregular band near the hinder margin; the under side is grey, tinged with green, having spots nearly corresponding to those above.

THYMELE TAGES. *Dingy Skipper.* Larger than the preceding; the surface rather dark brown, clouded with ash-grey, a few white points, and a series of them round the margin of all the wings; under side tawny grey, with ill-defined white spots. Caterpillar bright green, with a yellow stripe, dotted with black, down the back, and others on the sides. Not unfrequent at the end of the month.

SATURNIA.

Generic Distinctions.—Antennæ beautifully fringed, like the feather of a pen, in the male; *palpi* and *trunk* wanting; *head* small; *wings* very broad and entire.

SATURNIA PAVONIA MINOR. (Frontispiece.) *The Emperor Moth.* This beautiful species sometimes attains to three inches in width : the colour is greyish brown, faintly tinged with purple ; the hinder margin of all the wings has a band of pale brown, with two transverse bands of brown and purple on each wing, the hinder band very much waved ; the centre of each is also ornamented with a large ocellus, placed on a light ground, consisting of a black pupil with a yellow or grey iris, and partly surrounded by a light-blue crescent ; in the apex of the anterior wings there is a patch of purple and a black or whitish mark. The caterpillar is of a lovely green, and has a black band on each segment, adorned with pink tubercles, bearing a whorl of six hairs diverging like a star. Latreille mentions a manufactory established in Germany, in which the silk of which the cocoon consists is used ; the singular form of this cocoon was described in March.

LEPIDOPTERA. NOCTUIDÆ.

PLUSIA.

Generic Distinctions.—Antennæ simple ; *palpi* longer than the head ; *proboscis* long ; *head* and *thorax* tufted.

PLUSIA GAMMA. *Gamma Moth.* Upper wings grey, variegated with dusky brown, having a pale spot towards the apex, and a few transverse dark lines slightly waved ; the disc is inscribed with a silvery character resembling the letter G or the Greek *gamma;* the under wings are ash-brown ; the hinder margin and nervures dark brown ; head and thorax grey. Caterpillar light green, with faint yellow lines on the sides, and white ones on the back.

PLUSIA CHRYSITIS. (Plate XIV.) *Burnished-brass Moth.* Anterior wings ornamented with two broad golden-green bands, variable in their tint ; near the apex there is a transverse line of deep brown, of which colour are also the hinder wings. The caterpillars of this species are green, with a longitudinal white line on the sides, and oblique streaks of the same on the back. The Moth frequents lanes and the rank vegetation found among rubbish, and is common in the south of England.

LEPIDOPTERA. NOCTUIDÆ.

MISELIA.

Generic Distinctions.—Antennæ long, robust, and sometimes slightly serrated; *maxillæ*, length of the antennæ; *palpi* short; *head* clothed with scales; *thorax* large and crested; *wings* narrow.

MISELIA APRILIANA. *April Miselia.* The upper wings in this species are of a fine green, marked with transverse black lines and spots; the under wings are dusky brown, with a light streak on the inner side, and another on the hinder margin. The caterpillar is commonly ash-grey, with dark spots and lines, but varies considerably. It feeds on the ash, elm, and beech; there are three broods in the year.

LEPIDOPTERA. NOCTUIDÆ.

PHLOGOPHORA.

Generic Distinctions.—Antennæ long, slender, and simple; *palpi* ascending; *thorax* crested; *anterior wings* longitudinally folded in repose, deeply indented, and rather elongated.

PHLOGOPHORA METICULOSA. *Angle-shades.* Upper wings pinkish-white, clouded with olive-brown, each with a large triangular purplish mark in the centre, beyond which there is a white band, the margin marbled with olive-brown; the hinder wings are whitish, with a faint rosy tinge, having a dusky central crescent, and two or three faint waved lines. The caterpillar is green, with a row of white spots on the back, and a white line on each side; it feeds on culinary vegetables, and many common field-plants. There are three broods, which appear in April, June, and September.

NEUROPTERA. PSOCIDÆ.

ATROPUS.

Generic Distinctions.—Tarsi three-jointed; *wings* wanting.

Having mentioned the Death-watch, *Anobium*, it seems desirable to name another insect sometimes confounded with it, owing to a similarity in the sound it produces, viz.—

ATROPUS PULSATORIUM. This is a very minute insect, of a dirty-white colour, commonly found amongst old books and on papered walls. Its specific name alludes to the noise which it makes, similar to the ticking of a watch. The larvæ resemble the perfect insect both in habits and appearance.

.

CHAPTER V.

MAY.

MAY has ever been a favourite month; poets sing its praise, as uniting all the opening beauties of spring with the brightness and radiance of summer, as glowing with flowers, perfumed with sweet odours, and melodious with the cheerful carolling of birds. "Goddess of the spring" is only one among the many flattering epithets applied to it ; and the poet Darwin says,—

> "For thee the fragrant zephyrs blow,
> For thee descends the sunny shower ;
> The rills in softer murmur flow,
> And brighter blossoms gem the bower."

But the praises lavished so profusely were originally uttered in more southern climates, and in our northern latitude

May-day has but scanty garlands for her Queens and May-poles, even where this pretty and appropriate welcome has not died away with the other rural festivities of our more simple ancestors. Towards the latter end of the month, however, the country is profuse in beauty of various kinds ; many insects may be seen on the wing, sporting in the bright sunshine and enjoying their happy though brief existence. The lovely little blue Butterflies, *Polyommati*, may generally be seen in profusion, their small size, brilliant colour, and delicate markings, rendering them very attractive ; the Fritillaries begin to display their silver-spotted wings, and the Peacock Butterfly, so well known and universally admired, greets the eye of the lover of nature ; several Moths also make their appearance this month, tempted by its more genial atmosphere.

Amongst the Beetles may be seen species of the ex-tensive family of Cockchafers, *Melolonthidæ*, some of which are large and handsome : they live on the leaves of trees, for which the structure of the mouth is well adapted ; and in the larva state, which lasts three or four years, they do equal injury to the roots of various plants. The grubs of the common Cockchafer are white and fleshy ; at the com-mencement of spring they quit their winter retreats, at a great depth in the earth, and come within an inch of the surface ; when full-grown they again retire to the depth of two feet, where they become pupæ, having before constructed a cell, of an oval form, and very smooth in the inside. They assume the perfect state in February, but do not venture into the air until the fine days of May ; their existence is then but short ; remaining inactive during the day, they only emerge from their retreats at sunset, and fly humming round the trees. In favourable seasons they swarm to a great extent, and the damage committed by them is often serious. Mouffet informs us that in 1574 so great a number of these insects were driven into the Severn, that they hindered the mills from working, and it required the united efforts of hawks, ducks, and people to destroy them. Some districts in Ireland were completely devastated by them many years since ; they were first seen in Galway, hanging from the trees in clusters, dispersing at sunset " with a strange humming sound like the beating of drums, and darkening the air for the space of two or three miles square ; in a short time they entirely ate up all the leaves, stripping the trees as bare as in the depth of winter : the multitude spread so much that they infested the houses, and became extremely troublesome. Happily, high winds and rain at length checked them ; the pigs and poultry, too, watched beneath the trees, and were fattened by them; even the country-people, then labouring under a scarcity of provisions, had a way of dressing and living upon them as food."

A very pretty and attractive visitor is frequently seen in May, lighting the grass and herbage with its tiny green . lamp. The Glowworm, as this interesting insect has been named, may be seen from the latter end of this month till August; its scientific name is *Lampyris noctiluca;* the body is long, depressed, and of a soft consistence ; the antennæ rather short, and serrated ; the thorax nearly square, and concealing the head. It is the female that emits the light; the male, possessing wings, of which the female is destitute, is thus enabled to find its mate, who would otherwise be concealed from his view. Poets have frequently alluded to this pretty domestic lamp, but many have attributed it to the male ; thus Cowper says,—

> " This truth divine
> Is legible and plain :
> 'Tis power Almighty bids *him* shine,
> Nor bids *him* shine in vain."

Shakspeare also alludes to it in the following lines :—

"The Glowworm shows the matin to be near,
  And 'gins to pale *his* ineffectual fire."

Darwin, however, was more enlightened, as he thus describes
this beautiful light :—

"You,
Warm on her mossy couch the radiant worm,
Guard from cold dews *her* love-illumined form,
From leaf to leaf conduct the virgin light,
Star of the earth, and diamond of the night."

Some authors have endeavoured to disprove the idea that
the Glowworm's light is intended for this purpose, and
assert, in proof of their opinion, that the male is also luminous, though in a less degree ; this, however, does not seem
sufficient to refute the popular belief, and, until some other
use is discovered for this phenomenon, we cannot err in
ascribing to it the poetical one which has gained such universal belief. The eyes are very peculiar: Mr. Knapp says,
"When viewed at rest, no portion of the eyes is visible,
but the head is margined with a horny band, under which
the eyes are placed ; this prevents all upward vision, and
the blinds are so fixed at the sides of the eyes, as greatly to
impede the view of lateral objects ; the chief end of this
creature's nightly peregrinations being to seek his mate beneath him on the ground, this apparatus seems designed to

facilitate his search, by confining his sight entirely to what is below him, just as we place our hand over the brow to enable us to see more clearly an object on the ground." This may or may not be the cause of the peculiar formation, but similar proofs of contrivance are constantly presenting themselves to those who study the works of the benevolent Creator, who does not disdain to adapt the eye of a little insect to its wants and happiness.

An interesting account of these insects is given in the " Magazine of Natural History" for November, 1835. Mr. White, having collected some females at the end of June, confined them in a glass jar, with sand at the bottom, covered with moss; he supplied his captives with snails, on which they throve well, eating during a whole day without intermission, and then fasting for eight days. About the middle of July, they deposited their eggs in the moss, and soon after died: the eggs were of a pale yellow, and emitted light, particularly when the moss was sprinkled with water. In August the larvæ appeared : they were rather lighter in colour than the eggs, but became gradually darker; they had also the power of giving light. These larvæ went through the usual process of casting their skins, and in the following May (having been nine months in attaining their full growth) they changed into pupæ, and at the end of the month the perfect insects made their appearance. The larva of this Beetle very much resembles the female, but may be distinguished by its larger size, by the colour, which is black instead of brown, and by the imperfect structure of the legs and antennæ. We will now resume the description of species.

MELOLONTHA.

Generic Distinctions.—*Antennæ* of ten joints, produced into thin leaflets in the male; *body* oblong, and often hairy; *thorax* slightly convex; *elytra* shorter than the body.

MELOLONTHA VULGARIS, *Common Cockchafer*, has the antennæ and elytra of a reddish brown, the latter have four longitudinal ribs; the breast is grey and downy, and the margin of the body marked with a row of triangular white spots. The insect is well known under its English name of Cockchafer. The body is of a peculiar shape, being pointed behind; this, and the beautiful fan-like antennæ of the male, are both worthy of remark.

COLEOPTERA. CARABIDÆ.

NEBRIA.

Generic Distinctions.—*Antennæ* filiform, or setaceous; *head* narrower than the *thorax*, which is heart-shaped; *palpi* short; *body* oblong and depressed; *legs* long and slender; *colour* generally black or brown.

NEBRIA COMPLANATA. Buff or clay-colour, with black markings. Some of the species are very handsome, and are generally found on the sea-shore.

COLEOPTERA. HARPALIDÆ.

HARPALUS.

Generic Distinctions.—*Antennæ* filiform; *palpi* filiform; *elytra* entire, not truncated at the extremity; *thorax* rather square; *body* oval; *colour* generally shining black, with reddish limbs.

HARPALUS RUFICORNIS. Brownish black above, black below, with the antennæ and feet pale brown; about half an inch long; the elytra are downy, and lined or striated. The Beetles of the family *Harpalidæ* are generally found under stones, decaying leaves, and in similar situations; they run with great agility, and some species fly well. Some genera feed on small insects ; others prefer vegetable food, that named *Zabrus* being very destructive to wheat. The larvæ are longish, furnished with strong jaws and a forked tail.

EPHEMERA.

Generic Distinctions.—*Antennæ* very short, terminated by a seta; *mandibles* nearly obsolete; *palpi* very indistinct; *legs* long and slender, the first pair being inserted close to the head; *body* long ; *wings* triangular and horizontal.

EPHEMERA VULGATA (Plate IV.), *Common May-Fly*, is of a greenish-brown colour, having transparent wings, mottled with brown ; it has three long black bristles at the end of the body.

EPHEMERA ALBIPENNIS is remarkable for the whiteness of its wings, so that the swarms look like a fall of snow. The generic name is given in consequence of the short duration of the insect's life after attaining its perfect state. The *Ephemeræ* appear in swarms after sunset, from the latter end of the month till autumn, flying along the margins of streams, alternately rising and falling, with an

I

elegance of motion which cannot fail to attract attention. The larva inhabits water, remaining concealed in the day-time under stones, or in holes made in the clayey soil ; it is of course then destitute of wings ; the antennæ are long ; the mouth provided with a pair of horny appendages, like jaws ; and the body has a series of leaf-like plates on each side, with gills, or branchiæ, at their base ; the extremity of the body is furnished with short filaments. The pupa differs only in having the rudiments of wings.

HYMENOPTERA. ANDRENIDÆ.

ANDRENA.

Generic Distinctions.—*Antennæ,* third joint longer than the others ; *maxillæ* long and bent ; *labrum,* or *lip,* short ; *proboscis* downy and thick ; *hind legs* hairy.

ANDRENA NIGRO-ÆNEA. (Plate IV.) This species of Bee has the body black, and densely clothed with tawny-coloured hairs ; the antennæ are black, as is the thorax, which is also covered with reddish hair ; the wings are transparent and slightly iridescent; the legs are black and hairy, the hinder pair being thickly clothed with long white hair.

These insects make their nests in banks composed of light soil, and prefer a southern aspect. They excavate holes of a cylindrical form, nearly a foot in depth, of such a diameter as to admit the insect : in making these burrows, they remove the earth grain by grain, and throw it up outside, in the form of a hillock ; some species penetrate in a horizontal, others in a perpendicular direction. They construct a cell at the bottom of the hole, which they partly fill with pollen, made into a paste with honey; and in this they deposit their eggs, taking care to stop up the mouth of the hole carefully, to prevent the ingress of Ants or other insects which might injure the young larvæ. There are many other interesting genera in this numerous family which cannot be described here : they all consist of two kinds of individuals only, the males and females ; they are not therefore strictly social insects, like the Hive and Humble Bees, the true Wasps and Ants, all of which are provided with neuters, for the due support of the community.

HYMENOPTERA. VESPIDÆ.

VESPA.

Generic Distinctions.—*Antennæ* generally elbowed, and thickened

I 2

at the tip; *mandibles* short; *body* smooth and polished, generally black, marked with yellow; *wings* longitudinally folded.

VESPA VULGARIS, the *Common Wasp,* has the antennæ, head, and thorax black, marked with yellow; body yellow, with the base of the segments, and a spot on each, black.

VESPA CRABRO, the *Hornet,* is much larger than the common Wasp, of a rich brown, with dark markings; the head and body buff, spotted with brown.

The family *Vespidæ* is divided into *Vespa,* or true Wasps, and *Odynerus,* comprising the solitary species; of the former, the Hornet is the largest of the species, and its sting is a formidable instrument of defence. The nest of this insect is of similar construction to that of the common Wasp, though of coarser materials, and the columns supporting the rows of cells are much stronger; it is constructed either in the hollows of trees, the thatch of barns, or in timber-yards. It is difficult to obtain a sight of their nests while building; for should the aperture be too large, they erect a wall of the same material as the cells, which is described by some naturalists as decayed wood, by others, as the bark of trees gnawed to pieces and moistened with a sticky fluid, which the insects have the power of emitting. With this they make a kind of pasteboard, thicker than that of the

Wasp, but not so pliant, and of a dull buff-colour: if the nest does not fill the cavity in which it is commenced, they protect it by a thick piece of the same substance, made very similar in form to the paper cones in which grocers put their sugar. *Vespa vulgaris*, the common Wasp, makes its nest in the ground : early in the spring, a few large Wasps may be seen actively engaged in obtaining materials for this purpose ; these are females which have survived the winter, and are now preparing a nest for the purpose of depositing their eggs; they form a layer of hexagonal cells, in each of which an egg is placed, and the larva is hatched in a few days. The grubs are fed by the parent until full-grown, when the mouth of the cell is closed, and they become pupæ ; in this condition they remain about ten days, when the first brood appears in the winged state, and, being all neuters, they are ready to assist their parent in enlarging the nest, constructing fresh cells, and feeding the larvæ, as the foundress of the colony still continues to deposit her eggs. It is not till the latter end of the summer that male and female Wasps are hatched : the latter are not driven from the nest, as is the case with the Hive Bees ; consequently there is no swarming of Wasps, but all remain together till the cold of the advancing season destroys the numerous family, with the exception of about a dozen females, who survive to form fresh colonies in the ensuing spring. As the nest is now a desolate ruin, we will take the liberty of examining it without fearing the stings of its waspish inhabitants. A full-sized nest is nearly a foot in diameter, of a globular form; the outer covering is constructed of many layers of a thin paper-like substance made of moistened wood. There are several layers of cells, each about two-thirds of an inch in depth, and supported by numerous strong pillars; a nest consists of about 16,000 cells, differing in size according to the three orders which compose the family. These little insects feed on sweets of various kinds, as well as on flesh and insects, and frequently plunder hives to feed on the honey of its industrious inhabitants; one Wasp is said to be a match for three Bees, and they will boldly encounter a whole swarm to obtain their favourite food.

The solitary species of Wasp, forming the genus *Odynerus*, are much smaller than the common Wasp, which they resemble in colour; the females construct their nests in sandbanks, the crevices of walls, or in decayed wood. These holes are several inches in depth, and of a cylindrical form, the entrance being defended, in some species, by a curved

W Wang del et sc

1 Papilio Machaon.   2 Gonepteryx Rhamni   3 Pontia Brassicae

way, formed of sand. In the interior, the female buries eight or ten caterpillars, similar in species and size, arranging them in a spiral direction ; an egg is then deposited in the middle, and the mouth closed. The young Wasp, when hatched, devours the caterpillars, which are just sufficient for its support ; then assumes the pupa state, in a cocoon of slender texture ; and at length, becoming a perfect insect, flies away.

There are about twenty British species of this genus, but the specific characters are liable to considerable variation, which renders them difficult to investigate.

LEPIDOPTERA. PAPILIONIDÆ.

PAPILIO.

Generic Distinctions.—*Antennæ* rather long, with a slightly curved club; *palpi* short, third joint minute; *legs* all formed for walking; *hinder wings* scolloped, with a long narrow projection like a tail, and grooved to receive the body; *caterpillars* smooth and naked; *chrysalis* angular, and fixed by a band round the middle.

PAPILIO MACHAON. (Plate V.) *Swallow-tail.* This most beautiful insect is the largest of our Butterflies, the female sometimes measuring nearly four inches; the base of the upper wings is black, slightly tinged with yellow; the apex principally of the same colour, with a row of semicircular yellow spots; the margin is also edged with yellow, spotted with black, the latter forming three large patches, and also broadly marking the nervures; the base of the under wings is yellow, except the inner side, which is black, the nervures being dusky; beyond the yellow portion is a broad black band, marked by faint blue spots, and six large yellow crescents; the outer edge is yellow, and on the hinder angle is a large round spot of red, streaked with blue; the under side resembles the upper in all essential points. The caterpillar is greenish, with a black band on each segment, spotted with red; it feeds on umbelliferous plants. The perfect insect is rather local, but in some places tolerably abundant; it continues till August.

LEPIDOPTERA.   ERYCINIDÆ.

NEMEOBIUS.

Generic Distinctions.—*Antennæ* with a large abrupt club; *palpi* very short; *upper wings* triangular; *lower wings* rounded; anterior *legs* very short.

NEMEOBIUS LUCINA. *Duke of Burgundy Fritillary.* This, which is the only British species of this genus, measures about fourteen lines ; the upper surface is dark brown ; anterior wings with three series of light yellowish spots ; the marginal row with a black dot in each ; the posterior wings have also a row of yellow spots, with the margin similar to that of the upper pair ; the under side is much paler, marked with light spots and streaks of black ; the hinder wings have two pale bands, composed of oval marks, the outer one edged with black. The caterpillar is long, oval, and depressed, of a pale olive-brown, with a large black spot on each segment ; it feeds on the cowslip and primrose. The butterfly is rather local.

LEPIDOPTERA. PAPILIONIDÆ.

PONTIA.

Generic Distinctions.—See page 95.

PONTIA BRASSICÆ. (Plate V.) *Common Cabbage Butterfly.* The wings of this well-known insect are white above, with a large patch of black on the tip of the anterior pair ; the male has no other mark, excepting a black spot on the outer edge of the secondary wings ; but the female has in addition two others on the upper wings, and a patch of the same colour at their hinder margin ; on the under side the wings are yellowish, the upper pair having two conspicuous black spots on each ; the body and antennæ are black. The caterpillar is green, with a line of yellow on the back, and one on each side ; the body is covered with black tubercles, each with a hair in the centre. It consumes the cabbage, brocoli, cauliflower, etc., and has been found on the turnip : these vegetables would be entirely destroyed were it not for the many enemies which prevent their increase ; small birds devour immense numbers ; a titmouse will take half a dozen to its nest in a very short time. In enclosed gardens, sea-gulls with their wings cut are of infinite service, and poultry of any sort will soon clear the ground of these destructive little creatures.

PONTIA CHARICLEA. *Early White Cabbage.* This species differs from the last in size and rarity, having been observed in Hertfordshire and Derbyshire only.

PONTIA NAPI. *Green-veined White.* Colour yellowish-white, with the tip of the primary wings dusky, the male having one, and the female two black spots nearly in the middle of each ; the hinder pair are free from marks, except the ordinary dusky spot on the anterior margin ; on the under side, the nervures are strongly marked with a line of dingy green; the upper wing having two black spots near the hinder margin. The caterpillar is dull green, covered with white raised spots, blackish at the tip, and tufted with short hairs; the perfect insect is very common, and presents many variations in size and markings.

PONTIA SABELLICÆ. *Dusky-veined White.* This is considered by many writers to be merely a variety of the *P. napi,* and indeed the only difference seems to consist in the colour of the nervures.

LEPIDOPTERA. LYCÆNIDÆ.

POLYOMMATUS.

Generic Distinctions.—*Antennæ* short, with a compressed club; *palpi* a little longer than the head; *wings* entire; *colour* blue, but dissimilar, the females being generally brown; *size* small.

POLYOMMATUS ALEXIS. (Plate IX.) *Common Blue.* This very pretty insect is bright lilac-blue in the male, the hinder margin edged with black; the anterior edge of the upper wings white; the female is brown, powdered with blue towards the base of the wings; sometimes the surface is entirely purplish-blue, and ornamented with a darkish band, having ocelli on the hinder pair ; the under side in both is brownish ash-colour, the upper wings having two ocelli near the body, a slender streak, placed transversely, a row of ocelli near the middle, and externally a row of dusky crescents edged with reddish-yellow ; beyond these are dark spots on a white ground; the posterior wings have generally four ocelli near the base, an angular white spot near the middle, and a curved band of ocelli, succeeded by a series of markings similar to those on the superior wings. The caterpillar is hairy and of a green colour, subsisting chiefly on grasses.

POLYOMMATUS ARION. *Large Blue.* This species measures an inch and a half; the wings are pale violet-blue, with a broad dusky border round the hinder margin ; the male having a group of black spots on the disc of the upper pair, and the female a similar group on all the wings ; the under side is ash-colour, inclining to brown ; the anterior wings with two ocelli towards the base, an irregular band of them beyond the centre, and a centre of lunules, with a small white mark adjoining each ; the posterior wings bluish at the base, where there are four spots. This species is rare, and there are two or three others still more so, which on that account will not be described.

POLYOMMATUS ARGIOLUS. (Plate IX.) *Azure Blue.* An

abundant species, very similar to *P. Alexis*; delicate blue, tinged with lilac, the wings edged with black; the female has a broad dusky border on the primary wings and a series of spots of the same colour near the hinder edge of the secondary pair; beneath, the colour is grey, tinged with blue; the upper wings with a slender curved spot in the centre, and a band of streaks; the hinder wings nearly similar. The caterpillar is like that of the Common Blue, and is found on the holly. The insect is seen principally in the south; there is another flight in August.

POLYOMMATUS ALSUS. *Bedford Blue.* This is the smallest British Butterfly, seldom surpassing an inch in size; the surface is brown, slightly tinged with blue; the under side ash-colour, with a black crescent, edged with white, on each wing, and a transverse series of ocelli near the hinder margin; on the inferior wings this series is very irregular, and there are several scattered spots towards the base. This delicate little insect seems to occur, though not abundantly, in most parts of the kingdom.

POLYOMMATUS ADONIS. *Clifden Blue.* Expansion of the wings about fifteen lines; the surface of the male a most beautiful azure-blue; the hinder margin of all the wings edged with black; on the hinder side the colour is brownish-grey, the base being blue, and the markings nearly similar to *P. Corydon*, which appears in July. The female is brown above, glossed with blue, with a small black spot on the primary wing, and on the secondary pair a light streak with ocellated spots.

LEPIDOPTERA. PAPILIONIDÆ.

LEUCOPHASIA.

Generic Distinctions.—Antennæ, club abrupt and compressed; *palpi* short and flat, the basal joint large and cone-shaped, the second short and four-sided, the third minute and almost globular; *wings* very narrow and oval.

LEUCOPHASIA SINAPIS. (Plate VI.) *Wood White.* This is the smallest of our white Butterflies, and has somewhat the appearance of a Dragon Fly; the colour milk-white; the base of the wings dusky, and a large brown spot on the anterior pair at the tip ; the under side is faintly tinged with yellowish-green. The caterpillar is green, with a line of bright yellow on each side. It feeds on the *Ornithopus perpusillus* (bird's-foot trefoil) and *Lathyrus pratensis* (meadow vetchling). The insect is rather rare: I took only one specimen in Sussex.

1 Pontia Cardamines 2 Leucophasia Sinapis 3 Pieris Cratægi
4 Melitæa Euphrosyne

LEPIDOPTERA. NYMPHALIDÆ.

MELITÆA.

Generic Distinctions.—See page 97.

MELITÆA ARTEMIS. *Greasy Fritillary.* Colour deep yel-lowish-brown ; primary wings with black undulating lines and light yellow spots ; secondary pair with three bands, the middle one tawny-orange, with six small black spots; the others irregular, and of a light yellow; beneath, the primary wings glossy, paler than above, but the markings somewhat similar ; the hinder pair with three bands of pale yellow spots, edged with black; between the central and marginal band, the row of spots on the upper surface is distinctly marked, each surrounded with pale yellow. The caterpillar is black above, with spines of the same colour ; the under side yellow, and a row of very minute dots on the back and sides : it feeds on the plantain. The butterfly occurs prin-cipally in the south, and is said to be very abundant near Brighton ; it is occasionally found near Glasgow.

MELITÆA EUPHROSYNE. (Plate VI.) *Pearl-bordered Fritillary.* Wings of a yellowish-brown above, blackish at the base, and variegated with transverse spots of black, each wing having a row of black spots towards the apex, and a band of the same on the outer margin, which forms a triangular point on the inner side, enclosing a row of spots, the colour of the ground ; the primary wings are yellowish underneath, the black spots smaller than the corresponding ones on the surface, the tips light yellow ; the hinder wings have several large yellowish spots at the base, some of which are slightly glossed over with silver, the spaces between being red ; the latter colour forms a large spot in the middle, the space between this and the hinder margin being variegated with rust-brown and yellow, and a row of dark spots ; the hinder margin is adorned with triangular silvery spots, bounded by a deep black line; there is also a long silver spot near the centre, forming part of a band of yellowish colour. The caterpillar is black and spinose, with lines of orange along the back. It feeds on the violet. The butterfly seems pretty generally distributed, and appears again in August.

LEPIDOPTERA.   LYCÆNIDÆ.

THECLA.

Generic Distinctions.—*Antennæ* gradually thickening to the apex ; *palpi* with the terminal joint short, slender, and oval ; *upper wings* entire, the lower pair with one or two small appendages ; *colour* always brown above.

THECLA RUBI. (Plate IX.) *Green Hair-streak.* Expansion of the wings about an inch ; the surface of a uniform brown colour, the under side of a fine green, sometimes having a row of white spots on the secondary wings. The caterpillar is downy, light greenish-yellow, with a row of yellow dots on each side, and a white line above the feet : it feeds on the bramble, broom, and other plants. There are two broods, but this is not a very common species, though found plentifully in some parts of Scotland.

<hr>

LEPIDOPTERA. NYMPHALIDÆ.

VANESSA.

Generic Distinctions.—Antennæ with an oval club ; *palpi* nearly meeting so as to form a kind of beak ; *wings* angular, having projecting points on the hinder margin ; the *caterpillar* armed with long spines ; *chrysalis* angular, suspended by the tail.

VANESSA Io. *Peacock Butterfly.* The colour of this beautiful and well-known insect is deep brownish-red, with a large eye-like spot, or ocellus, on each wing : that on the upper pair has a large yellow crescent on the inner side, a patch of blue externally, and a large reddish pupil ; there are also five white spots, and alternate patches of black and

K

yellow form the rest of the margin, near the base ; the border is yellow, with lines of black. The ocellus of the hinder wings consists of a black centre spotted with blue, and encircled with a zone of pale brown, bounded by a large black crescent ; the under side is dark brown, with waving lines of black, and a few white points. The caterpillar is of a shining black, with many white spots disposed in transverse lines : it feeds on the stinging nettle (*Urtica dioica*).

LEPIDOPTERA. LYCÆNIDÆ.

LYCÆNA.

Generic Distinctions.—*Antennæ* slender, the club thick and abrupt; *palpi* longer than the head ; *secondary wings* nearly straight on the inner edge, and slightly notched at the extremity; *colour* brilliant copper.

LYCÆNA PHLEAS. (Plate IX.) *Common Copper.* Upper wings copper-colour, spotted with black, the under pair brownish-black, with a copper band dotted with black on its outer edge; the under side of the primary wings spotted similarly to the upper, but the colour paler ; the secondary wings fawn-colour, with many indistinct marks, and a tawny band.

Plate IX

1 Hipparchia Ægeria . 2. Thecla Quercus .   3 Thecla Rubi . 4 Lycæna Phlæas .
5 Polyommatus argiolus   6 Polyommatus Alexis

This species is liable to many variations in the markings : specimens have been found in which the copper-coloured parts were nearly pure white. The caterpillar is said to be green, with a yellow stripe on the back : it feeds on the sorrel.

PAMPHILA.

Generic Distinctions.—Antennæ terminated by a hook, short, with the club straight, abrupt, and spindle-shaped ; *palpi* short ; *anterior wings* rather longer than in the genus *Thymele ;* the hinder pair have a small projection.

PAMPHILA PANISCUS. *Chequered Skipper.* Measures about an inch ; the surface is brownish-black, marked with . numerous bright, rich, brown spots ; those on the anterior wings consisting of a large patch in the middle, then an irregular curved band, intersected with black nervures, and lastly, a faint row of dots parallel with the hinder margin ; on the secondary wings are three spots, one of them larger than the others, and a band of small dots ; the under side is yellow, inclining to grey ; the upper wings have several marks, and the under pair seven round yellowish-white

K 2

spots, and a band of the same colour. The caterpillar is dark-brown, ornamented with two longitudinal yellow stripes; the head black, the segment behind it has an orange stripe: it feeds on the plantain. This pretty insect is rather local.

PAMPHILA SYLVANUS. *Large Skipper.* Colour dingy yellow, with the nervures and extremity of the wings brown, the latter deepening into a blackish line round the outer edge; the anterior wings have a few small yellowish spots on the dusky ground towards the tip, and an oblique black streak in the centre; the hinder pair are marked with indistinct cloudy spots of a yellow hue, which colour is sometimes diffused over a large portion of the disc; on the under side the anterior wings are yellow, and the posterior pair yellowish-green, the former dusky black at the base, and the latter marked with a curved series of pale spots, varying in size and colour. This insect may be found on the borders of woods from May till August.

LEPIDOPTERA. ANTHROCERIDÆ.

INO.

Generic Distinctions.—*Antennæ* slightly curved, and gradually thickening to the apex, pectinated in the male, in the female serrated; *palpi* short, and clothed with hairs.

INO STATICES. *Green Forester.* The upper surface of the superior wings and body are of a beautiful golden-green with a silky gloss, inclining in some parts to blue; the remainder of the upper, and the whole of the under side, are brown. The caterpillar tapers at both ends; the colour is green, with the head black, two rows of black spots on the back, and on each side a series of red dots; it feeds on the *Cardamine pratensis.*

This insect, though somewhat local, is of frequent occurrence near London.

LEPIDOPTERA. SPHINGIDÆ.

METOPSILUS.

Generic Distinctions.—*Antennæ* slightly thickening from the base; *anterior wings* very acute at the apex, with a slight curve on the hinder margin below the tip; the inner margin is also curved.

METOPSILUS ELPENOR. *Elephant Hawk-Moth.* The upper wings, which measure above two inches, are olive-brown, with the anterior edge, two oblique bands, and the hinder margin rose-colour; the hinder wings are dusky at the base, with the rest rose-colour ; the body is olive, with numerous stripes of deep rose.   The caterpillar, when full-grown, is brown, with six oblique stripes of a grey colour ; it frequents the different kinds of willow-herb and the common vine, and occurs frequently in many parts of England.

METOPSILUS PORCELLUS. *Small Elephant Hawk-Moth.* This species is much smaller than the preceding; the upper wings are principally ochre-yellow, marked with purple, the outer extremity having a purple band ; the under pair are blackish at the base, and purple behind, the intermediate space yellowish ; body deep rose-colour, or purple.   The caterpillar resembles that of *M. Elpenor.*

LEPIDOPTERA. SPHINGIDÆ.

SESIA.

Generic Distinctions.—*Antennæ* thickening from the base, and terminating in a seta; *proboscis* very long and spiral; *body* ending in a tuft of hairs, and of a short robust form ; *wings* clear and transparent.

SESIA FUCIFORMIS. (Plate XI.) *Brood-bordered Bee Hawk-Moth.* Body yellowish or olive-green, with the third and fourth segments deep red, and the two following yellow, the tuft of hair at the extremity black and yellow; the wings are transparent and iridescent, with the nervures, a band round the margin, and a streak on the upper pair purplish-brown; the base is tinged with green. The caterpillar is pale green, with the under side and the horn (the character-istic mark of the Hawk-Moth) reddish-brown: it feeds on the honeysuckle.

LEPIDOPTERA. SPHINGIDÆ.

SMERINTHUS.

Generic Distinctions.—*Antennæ* serrated; *proboscis* very short; *anterior wings* angular, and toothed on the margin.

SMERINTHUS TILIÆ. (Plate X.) *Lime Hawk-Moth.* This is a very variable insect, both in colour and in the form of the markings. The anterior wings are generally greyish, with an interrupted band of olive-green in the middle; the outer margin has a broad band of the same colour; the hinder wings are grey, with an ill-defined brown band running across the outer margin. At other times the insect is found of an obscure red; the thorax is marked with three longitudinal lines of olive-green. The caterpillar is pale green, with seven oblique whitish stripes on each side, edged with red or yellow: it feeds on the lime (*Tilia Europœa*). The insect is rare.

ARCTIA.

Generic Distinctions.—*Antennæ* rather long, bipectinate (having a double series of hairs) in the male, serrated in the female; *proboscis* very short; *palpi*, basal joint longer than the second; *wings* very thickly covered with scales; *colours* deep black, crimson and yellow, in spots and bars.

ARCTIA VILLICA. (Frontispiece.) *Cream-spot Tiger-Moth.* This species measures about two inches and a half; the anterior wings are deep black, spotted with cream-colour; under wings rather deep yellow, with spots of black, and a large black patch on the outer angle spotted with the ground-colour: the thorax is black, with two cream-coloured spots; the body yellow at the base, and terminating in red, with longitudinal rows of black spots. The caterpillar, when full-grown, is black, with tufts of greyish-brown hairs; it feeds on many common field-plants. This insect is not so common as others of the genus.

TRIPHÆNA.

Generic Distinctions.—Antennæ simple ; *palpi* short, and rising up in front of the head ; *proboscis* long and spiral ; *colour* bright, the under wings being always light yellow, with a black margin ; *size* moderate.

TRIPHÆNA FIMBRIA. *Broad-bordered Yellow Under-wing.* The head, thorax, and anterior wings are liver-colour, inclining to grey ; the latter have four pale transverse lines, and two pale rings on the disc, the second and third lines enclosing a space darker than the rest, in which the two ringed spots are placed ; the under wings are light yellow, with a broad band of black, edged with the ground-colour. The caterpillar is of an ochre-yellow, with a pale line along the back ; it feeds on the potato, violet, primrose, etc.

LEPIDOPTERA. LITHOSIIDÆ.

CALLIMORPHIA.

Generic Distinctions.—Antennæ slender and setaceous ; *palpi*, basal joint the same length as the two following ; *proboscis* longer than the head ; *anterior wings* long and narrow.

CALLIMORPHIA JACOBÆÆ. *Cinnabar Moth.* Anterior wings greyish-black, with a stripe of pink extending from the base nearly to the apex ; on the hinder border are two rounded spots, and a dash of the same colour ; the under wings are entirely pink, excepting a stripe of grey on the anterior edge. The caterpillar is yellow, ringed with black ; it feeds on the *Senecio Jacobæa*, or ragwort. Not a very abundant species.

DIPTERA. TIPULIDÆ.

TIPULA.

Generic Distinctions.—Antennæ simple, of thirteen joints ; *rostrum* long and narrow ; *palpi* short ; *body* elongated ; *wings* and *legs* both long.

TIPULA PRATENSIS. Body black, with the front and spots on the thorax reddish-brown.

TIPULA LUNATA. Reddish, with a black line on the upper part of the body ; wings of the same colour as the body.

TIPULA OLERACEA. Greyish-brown, without spots, the wings bordered with brown.

These insects are very commonly found in meadows,

where they may be seen rising in swarms, their long legs serving as stilts ; the females deposit their eggs in the earth at a short depth below the surface. The larvæ are fleshy grubs, which attack the roots of grass and other plants, doing much injury to the crops ; there are nearly fifty British species, which are well known under the name of Crane-flies, Harry Long-legs, etc.

DIPTERA. HIPPOBOSCIDÆ.

HIPPOBOSCA.

Generic Distinctions.—*Antennæ* inserted near the mouth ; *head* small, round, and attached to the thorax by a neck ; *thorax* large ; *wings* large and horizontal ; *body* soft ; *feet* short.

HIPPOBOSCA EQUINA. *Forest* or *Horse-Fly.* This insect has the head yellow and flattened ; the body, which is broad and short, yellowish, with brown spots ; the wings white, transparent, much longer than the body, and rounded at the extremity ; the body slightly hairy. This is a very troublesome species, living principally on horses, and abounding in the New Forest, Hampshire. It is a singular circumstance that the female Fly nourishes her young within her body, where the larvæ attain their full size, and even assume the pupa state, in which form they are deposited by the parent; this egg-like cocoon is at first soft and white, but soon hardens, becomes brown and of a round shape. These curious particulars were discovered by Réaumur, who, being anxious to observe the hatching of these singular eggs, as he thought them, carried some in his pocket by day, and took them to bed with him at night, that they might have the proper warmth; his surprise was great, when, instead of grubs, perfect Flies were produced. Mr. Curtis observes that "these flies move swiftly, and, like a crab, sideways or backwards, and that they are very tenacious of life." It is also remarked by Latreille that the ass fears them the most, but that they cannot cause much pain, or horses could not live in forests in the summer.

141

CHAPTER VI.

JUNE.

In this variable climate, June may generally claim with more justice those honours which the poets have accorded to May, and may well be considered the loveliest month of the year; the weather is less capricious, the trees are in their freshest robes, a profusion of the sweetest flowers is scattered over the ground, and numberless insects are called into being by the increasing heat, affording a never-failing source of amusement and instruction to the naturalist. The young student must now lose no opportunity of increasing his collection, and adding to the knowledge he may have already acquired of the habits and forms of the insect world: every advance in this knowledge will prove an increase of pure and lasting pleasure, from the tendency it must have, when properly directed, to raise the mind to the contemplation of the great Author of all things. June is particularly prolific in specimens of the order *Lepidoptera*, as might be expected from its sunshine and its flowers; their light forms and brilliant colours adding another charm to this gay and delightful season; the *Polyommati* in their delicate blue dress, the *Hipparchias* of more sober hue, and the *Fritillaries*, will, I hope, be in some measure recognized as old favourites, for many species will present themselves to our notice, as well as several new genera, both of diurnal and night-flying *Lepidoptera*. Some of the Hawk-Moths are particularly splendid in colour, and remarkable in size and form; whilst the more delicate beauty of the true Moths is not less worthy of admiration. The order *Coleoptera* will enrich the cabinet of the zealous collector with specimens both interesting and beautiful. One species of the family *Lucanidæ* makes its appearance, and will attract attention by the formidable length of its mandibles, bearing a resemblance to the antlers of a deer, to which it owes its name of Stag-Beetle. The family *Curculionidæ* is noted for the destruction caused by the larvæ of many of its species to our vegetable productions: they may be distinguished by their frontal elongation. The

Cetonidæ will please by their elegance, and the brilliancy of their colours, in which green usually predominates ; the English species, *Cetonia aurata*, or Rose-chafer, is probably known to many of my readers : it frequents flowers.   The *Silphidæ*, though less striking in form and colour, are interesting from their habits, which render this family eminently serviceable to man, and amply atone for their inferiority in beauty.   None of our British *Coleoptera* indeed can vie with those of tropical climates in the richness of their colouring; some of those are so splendid as to be worn instead of precious stones, the brilliancy of their hues being often greatly enhanced by a high degree of lustre, and diversified markings.   To many tropical Beetles might be applied the words of the poet, who is thus describing the birds which ornament the glowing landscape :—

> " With their rich restless wings, that gleam
> Variously in the crimson beam
> Of the warm west—as if inlaid
> With brilliants from the mine, or made
> Of rainbows."

Brazil is the richest country in the world in *Coleoptera*, and a recent French writer says, that " in the middle of January they are seen in the greatest profusion; the herbaceous plants are covered with brilliant beetles ; and the slender twigs of the *Mimosa*, on which they live in society, appear to bend under the weight of diamond beetles ; the *Lampyridæ*, issuing in myriads from their retreats, diffuse their mild effulgence over the plants and shrubs, which they often cover with their numbers, and the luminous *Elateridæ* dart about in all directions, filling the air with their radiant tracks."

In the Old World, the western coast of Africa, the Cape of Good Hope, Java, and other islands produce the greatest numbers of these brilliant insects; but leaving them to those who can search for their beautiful and extraordinary forms in their native lands, we must be satisfied with the more moderate degree of lustre in the tribes we possess, in which will be found no want of beauty and interest. The numerous family *Muscidæ*, so well known under the general name of Flies, present themselves abundantly to our notice in this and the following month ; the Domestic Fly, Blow-Fly, Blue-bottle, etc., are familiar to every one ; they frequent houses, woods, hedges, and, in fact, may be seen everywhere; they belong to the order *Diptera*, and some idea may be obtained of the great number of genera and species, by the fact that eight hundred quarto pages have been written by one author, comprising technical descriptions of a portion only; one genus contains 315, and another 230 species. The legs of the *Muscidæ*, like those of other Dipterous insects, are terminated by a tarsus consisting of five joints, the last being armed with two claws and furnished with lobes, which enable the insect to perform the curious, though common, feat of walking with the back downwards on the ceiling, and on highly-polished substances. It has been generally supposed that this was effected by the formation of a vacuum, caused by the close application of the margin of these lobes, and the muscular rising of the central parts; but Mr. Black-wall, in a volume of the Linnæan Transactions, has suggested a new solution of this interesting point. He says, that after much research into the subject, and many experiments, he breathed into phials containing Flies and other insects capable of walking in this manner, and found that when the moisture was condensed on the surface, it totally prevented the insects from taking hold of the glass; the same ensued when a little oil was substituted, or flour, pulverized chalk, etc., which adhered to the lobes of the foot. These facts seem to imply that an adhesive secretion is emitted by the instruments employed in climbing, and this, by the aid of highly magnifying powers, was found to

L

be the case. These interesting questions will, no doubt, be
the subject of further research, and every inquiry is sure to
open fresh sources of wonder and admiration.

COLEOPTERA. SILPHIDÆ.

NECROPHORUS.

Generic Distinctions.—*Antennæ* but little longer than the head,
the four last joints being perfoliated (not closely applied to each
other); *mandibles* not toothed; *elytra* of an oblong square form,
leaving the last three or four segments of the body uncovered;
colour dark brown, varied with yellow.

NECROPHORUS HUMATOR. (Plate II.) This species is
brownish-black, with the three last joints of the antennæ
orange-yellow; the elytra are deeply punctured, and each has
three slightly elevated lines ; the breast, and also the legs,
are covered with yellowish hairs. This insect is of frequent
occurrence in England, and all the species are designated
by the name of Burying Beetles, from their habit of interring
small animals after death, as a receptacle for their eggs,
thereby fulfilling a very important office in the economy of
nature. Any small dead animal, such as a mole or a mouse,
is soon visited by the *Silphidæ*, who creep beneath the body,

and commence scratching away the earth till they have made
a pit, into which the animal gradually sinks ; when it has
reached a sufficient depth, the earth is thrown over it, and
the young larvæ, when hatched, find themselves in the midst
of a repast, disgusting enough, but most happily suited to
their taste ; for the ground is thus freed from putrid substances, which would otherwise affect the purity of the
atmosphere.

————

GEOTRUPES.

Generic Distinctions.—Antennæ with a club divided into laminæ ;
mandibles standing out from the head, and notched; *thorax* very
convex, and as broad as the elytra, which are short and oval.

GEOTRUPES STERCORARIUS is entirely black above, tinted
with violet; the thorax is without punctures on the disc,
but has a few at the sides, and a short line in the middle ;
the elytra are marked with deep grooves, the spaces between
being smooth and convex; the under side and legs are steel-
blue, beautifully glossed with purple or green.

It is to this insect that Shakspeare alludes under the name
of the "shard-born Beetle," both in ' Macbeth' and ' Cym-

beline.' These insects seem to prefer still, dull evenings for their flight, at which time the humming noise they make is very considerable.

LUCANUS.

Generic Distinctions.—Antennæ, the four terminal joints project on one side; *head* as wide as the thorax; *maxillæ* terminating in a slender lobe; *mandibles* very large.

LUCANUS CERVUS. (Plate II.) The male of the Stag-Beetle is about two inches in length, including the mandibles; it is entirely of a brownish-black colour, the surface shining and covered with small punctures; near the fore leg is a patch of golden-coloured hair, which seems to be used for the purpose of cleaning the antennæ. The female is considerably smaller than the male; the mandibles are also short and the head much smaller.

The very formidable mandibles of this Beetle are employed in wounding the bark of trees, in order to feed on the sap. Mr. Waterhouse kept one alive for many weeks, feeding it on sugar and water; it also seemed fond of the juice of raspberries and other sweet substances. The perfect insect is found on the trunk of the oak, elm, and willow, and appears generally towards the middle of summer.

CETONIA.

Generic Distinctions.—*Antennæ* ten-jointed, the club with three joints; *thorax* widening towards the hinder margin; *body* nearly ovate, obtuse, and somewhat depressed.

CETONIA AURATA. This pretty insect, known under the name of Rose-chafer, is one of the most beautiful of British Beetles; the colour is a fine golden green, very shining above, and a bright coppery hue beneath; the elytra are ornamented near the tips with numerous transverse white marks; the form of the body is rather obtuse. This insect is found abundantly during the summer months, particularly in gardens, flying well, and with a humming noise, in the heat of the day.

RHYNCHITES.

Generic Distinctions.—*Head* inserted far into the thorax; *rostrum* enlarged at the extremity; *body* nearly square.

RHYNCHITES POPULI. (Plate II.) In this well-known species the body is shining, of a golden green, or with a bluish tint on the upper side, and dark violet beneath; the antennæ are black; the elytra punctured; in one of the sexes there is an acute spine on each side of the thorax, projecting forwards. The length of this insect is three lines; it is found on poplar and birch trees.

RHYNCHITES PUBESCENS is somewhat longer in proportion than the preceding, of a deep violet-colour, and clothed with rather long hairs; the elytra are marked with punctured lines.

RHYNCHITES BACCHUS. A beautiful species, found chiefly in Kent on the vine, and doing great injury by devouring the tender shoots, extracting the sap with its long tubular proboscis: this causes the leaves to roll up, and in these little tents, surrounded by a silken covering, the eggs are deposited.

The species of the family *Curculionidæ* have the English name of Weevil, and may always be known by the elongation of the head into a snout or rostrum.

COLEOPTERA. SILPHIDÆ.

NECRODES.

Generic Distinctions.—Antennæ longer than the head, thickening from the fifth joint to the apex; *thorax* nearly round.

NECRODES LITTORALIS is entirely black, with the three terminal joints of the antennæ yellow; there are three elevated lines on each elytron, the spaces between being thickly punctured; the hind legs are very thick, and toothed on the under side. It is found on the sea-shore and banks of rivers, under seaweed and carrion, on the latter of which it feeds.

———

COLEOPTERA. DERMESTIDÆ.

DERMESTES.

Generic Distinctions.—Antennæ short and clavate; *body* oval and somewhat convex; *legs* short.

DERMESTES LARDARIUS is of a dirty ash-colour, with three small black spots; it infests bacon, and may be found nearly throughout the year. *D. murinus* is generally found in moles, hawks, and other carrion, hung up against walls; and *D. vulpinus* infests the skins brought from Brazil to so great an extent, that it is said £10,000 have been offered for its destruction.

FORFICULA.

Generic Distinctions. — *Antennæ* filiform, of twelve or thirteen joints; *palpi* filiform; *tarsi* three-jointed; *wings* like a fan, and folded under two short elytra; *body* terminated by two scaly pieces forming a pincer.

FORFICULA AURICULARIA, the *Earwig*, has the body long, and of a reddish-brown, the eyes black, the thorax dark in the middle, and the sides yellowish, forceps yellowish-brown. Despised and disliked as are these harmless insects, owing to the generally received opinion (false as it is) of their creeping through the ear into the head, they are interesting on many accounts.   Usually, the care taken by female insects in constructing nests for the reception of their young appears more decidedly than in other animals the result of mere instinct, as the parent dies long before the birth of the young larvæ; but this is not the case with the Earwig, which seems gifted with something approaching to the maternal attachment evinced by the higher orders of creation, not only in taking great care of her eggs, but in brooding over them like a hen, and collecting them when scattered about. The young ones differ only from the parent in their small size, and the want of wings and elytra, which make their appearance in the pupa state; the wings are particularly beautiful, not only for the delicacy of their structure, but from the singularity of the nervures. Ought we, then, to feel contempt or dislike for an inoffensive little creature, the peculiarities in whose habits and structure testify that the hand of Omnipotence has been engaged in its construction? Why should it thus differ, unless to excite our attention and reward our research? The old naturalist Mouffet gives an amusing account of the destructive habits of the Earwig, which, though perfectly harmless to our persons, is mischievous in our gardens. "The English women," he says, "hate them exceedingly, because of the flowers of clove-gilliflower that they eat and spoil, and they set snares for them thus: they set in the most void places ox hoofs, hogs' hoofs, or old cast things that are hollow, upon a staff fastened into the ground, and these are easily stuffed with straw; and when by night the savages creep into them to avoid the rain, or hide themselves in the morning, these old cast things being shook, forth a great

multitude fall, and are killed by treading upon them."
They feed on vegetable substances, devouring not only fruit
but the petals of flowers.

TRICHOPTERA.   PHRYGANIDÆ.

PHRYGANEA.

Generic Distinctions.—Antennæ long and setaceous; *head* small;
mouth nearly obsolete; *inferior wings* larger than the superior, folded
longitudinally when at rest; *tarsi* five-jointed; *legs* spurred.

PHRYGANEA GRANDIS (Plate IV.) has the upper wings
brownish-grey, with a longitudinal black ray, and two or
three white points at the extremity.

The insects of this genus are found near water, and are
called Caddice-Flies or Water-Moths, and the larvæ,
Caddice-worms, in which state they reside in the water in
cases made of sand, shells, etc.; the perfect insects are
generally of obscure dark colours.

HIPPARCHIA.

Generic Distinctions.—See page 98.

HIPPARCHIA GALATHEA. *Marbled White.* The colours of this pretty Butterfly are black and greenish-white, in almost equal proportions; the upper wings have a large oval spot of the light colour near the base, and others, divided·by the nervures, in the middle, two small ones near the anterior angle, and a series of spots parallel to the margin; the lower wings are variegated in a similar manner, and have a row of triangular marks on the hinder margin; the under side is much paler than the upper, the light colour greatly predominating; there is also an irregular row of ocelli on the inferior wings near the edge, and an ocellus near the tip of the superior pair. These markings vary in form and colour; sometimes the black predominates, and in other specimens the light colour, but not so greatly as to prevent the insect being known. The caterpillar is yellowish-green, with a dark line on the back, and another on each side; the head is brown, and there are two small spines of the same colour on the hinder extremity; it is found in May on the *Phleum pratense,* or cat's-tail grass.   The Butterfly is usually seen in moist glades or in marshy grounds, and is not uncommon in the south.

HIPPARCHIA TITHONUS. *Large Heath.* Surface of the wings ochre-red, with the base and a broad margin dark brown; the upper wings have a large black spot with two minute white points, and the under pair a small ocellus just above the dark band. On the under side, the primary wings are similarly coloured to the upper, but the hinder pair are greyish-brown, with an irregular light band across the middle, in which there are usually four minute white points, surrounded with brown. The male is more deeply coloured, and has a brown cloudy mark on the primary wings. The caterpillar is green, with a reddish line on each side, and feeds on the *Poa annua*, or meadow-grass. The perfect insect is small, and of frequent occurrence.

HIPPARCHIA PAMPHILUS. *Small Heath.* This species is still smaller than the preceding, only measuring thirteen lines ; the colour of the upper side is light ochre-yellow, with the outer margins dusky, and surrounded by a fringe of whitish hairs ; the primary wings on the under side are ash-coloured at the tip, and ornamented with a black ocellus, having a white pupil ; the secondary wings are greenish-brown at the base, the rest ash-coloured and brownish-grey, with a few yellowish minute spots; these are, however, often very indistinct. The caterpillar is greenish, dusky on the back, and having a white line on each side; it feeds on the *Cynosurus cristatus*, or crested dog's-tail grass. The Butterfly is very generally found until September.

HIPPARCHIA JANIRA. *Meadow Brown.* This species is much larger than the two preceding, the wings of the female measuring nearly two inches; the ground-colour is brown, usually darkest in the male, and with a small ocellus near the apex, encircled with reddish-yellow as a rim, but in the female there is a large patch of ochre-yellow, within which the ocellus is placed; the hinder wings of the male are usually unspotted, but those of the female have an obscure yellowish mark in the centre; the under side of the primary wings is tawny-orange, with a broad pale band, in which is placed an ocellus; the hinder wings are dusky brown at the base and margin, the centre being paler, and sometimes marked with two or three minute black spots. The caterpillar is light green, with a white line on each side; it feeds on several kinds of grass. This is one of the most common Butterflies. Even in those damp cheerless summers, when few other insects are found, this hardy little creature may frequently be seen flying from flower to flower:

they were particularly abundant, it is said, in the arid summer of 1826.

HIPPARCHIA HERO. *Silver Ringlet.* Surface pale brown, inclining to yellow; upper wings pale, with an orange stripe near the hinder margin, and two small ocelli; the inferior wings with an orange line close to the edge, and four large ocelli; on the under side, the primary wings are orange at the hinder margin, which is adorned with a silvery line and two small ocelli; the secondary pair have a broad band of orange enclosing five large and two small ocelli, forming a curved band, behind which is a stripe of silver. This insect is rare.

LEPIDOPTERA. NYMPHALIDÆ.
MELITÆA.

Generic Distinctions.—See page 97.

MELITÆA CINXIA. *Glanville Fritillary.* This species is tawny orange-colour above, and the whole surface is reticulated (net-like) and spotted with black; the hinder wings having in addition a series of five or six black spots, parallel with the margin, but at some distance from it. Beneath, the colour is much paler, the primary wings have a few transverse black lines, and a series of black crescents towards the tip; the hinder pair have three irregular bands of pale spots, edged with black, and a row of black spots with an iris of ochre. The caterpillar is black, with white dots along the sides; it feeds on the *Plantago lanceolata*, or narrow-leaved plantain, and the *Veronica Chamædrys*, or germander speedwell. The eggs are hatched in the autumn, and the larvæ pass the winter in colonies, forming a kind of tent, by drawing together the leaves of the plant on which they feed, and covering the whole with a web of silk. The perfect insect is small and rather rare, though it has been found commonly in the Isle of Wight, and some places in the south. I have seen it (but rarely) in Sussex.

LEPIDOPTERA. NYMPHALIDÆ.

ARGYNNIS.

Generic Distinctions.—Antennæ rather long and slender, with a very abrupt spoon-shaped club; *palpi*, middle joint very long, basal and terminal joint short; *wings* very large, and slightly scolloped, hinder pair extending beyond the body.

ARGYNNIS ADIPPE. (Plate VII.) *High-brown Fritillary,*

In this species the wings often measure two inches and a half, and are of a yellowish-brown, with transverse undulating streaks, and round spots of black, also a series of black crescents near the outer margin; on the under side the primary wings are of a lighter hue, with a few silvery spots near the tip; the secondary wings yellowish-brown, with many silvery spots, some near the base, the rest forming two transverse bands; between the central and external bands there is a series of small round rusty-brown spots, most of them having a silver pupil. The caterpillar is reddish, but becomes olive-green with age, and has a white dorsal line and white dots on the sides; it feeds on the *Viola odorata*, or sweet violet. The insect is not uncommon in the south of England.

ARGYNNIS AGLAIA. *Dark-green Fritillary.* Very similar to the preceding on the upper side, but rather paler; the under side is green, particularly the inferior wings, which have six or seven silvery marks near the base, a curved band of the same near the middle, and another parallel with the hinder margin, consisting of seven spots without any intervening row of ocelli, as in the preceding species; the anterior wings have silvery spots at the hinder margin. The caterpillar is black, with red spots on the sides; it

feeds on the *Viola canina*, or dog violet. The perfect
insect is more plentiful than *A. Adippe*.

VANESSA.

Generic Distinctions.—See page 129.

VANESSA URTICÆ. (Plate VII.) *Small Tortoise-shell.* The
prevailing colour is orange-red, on the upper side of this
pretty insect; on the anterior margin it inclines to yellow,
and is marked with three large quadrate black spots, beyond
which is a small white mark; near the hinder margin is
another large black spot, and two small ones on the disc;
the margin is deeply edged with black, and ornamented with
a series of blue crescents, and two undulating lines of pale
yellow; half of the under wings is black, the rest orange-
red, with a border similar to that of the anterior pair; on
the under side, these wings are pale yellow, with three
patches of brown, the tip also mottled with the same colour,
and the border appearing to shine through from the upper
side; the hinder wings have the same appearance, being
brown, where the upper part is black, and of a lighter hue,

M

where it is so on the other side.   The caterpillar is blackish, with four yellow stripes, and beset with strong spines; it feeds on the nettle, and, when young, lives in families, but disperses when increasing size renders more food necessary. The butterfly is very common, and may be met with in March, but these are individuals which have survived the winter: there is another brood in September.

VANESSA C-ALBUM.   *Comma Butterfly.* This is the smallest species of the genus, and differs much in the form of the wings; the colour of the upper side is reddish-yellow, irregularly spotted with black, and having the hinder margin dark brown; two of the largest spots on the upper wings are placed on the anterior border, the others, four or five in number, occupy the disc; on the inferior wings there are usually three black patches; the colour of the under side is dark brown, often inclining to yellow, the basal half of the wings and the apex being the darkest; on the paler portion of the anterior pair is an obscure band of green, and two rows of greenish ocelli across both wings: the hinder wings have a short curved line of pure white near the middle, like the letter C reversed, hence the name of the species.   The caterpillar is brownish-red, with a broad band of white on the back; the head has two hairy tubercles, resembling ears ; it lives on the elm, willow, and various trees and shrubs. The butterfly is not generally common ; there is a second brood in September.

LEPIDOPTERA. LYCÆNIDÆ.

POLYOMMATUS.

Generic Distinctions.—See page 123.

POLYOMMATUS AGRESTIS. *Brown Argus.* This species measures about an inch ; the colour is dark brown, with a fine silky gloss, having a band of orange spots, and the primary pair a small black spot in the centre ; the under side is greyish-brown, with numerous ocellated spots, and a band similar to that on the surface ; the hinder wings have a white mark on the disc. It occurs in some abundance in the southern counties, appearing again in August.

LEPIDOPTERA. PAPILIONIDÆ.

PIERIS.

Generic Distinctions.—*Antennæ* slender, the club formed gradually; *palpi*, two lower joints robust, the basal one twice the length of the

M 2

second, the terminal joint very slender; *wings* very sparingly clothed with scales.

PIERIS CRATÆGI. (Plate VI.) *Black-veined White,* or *Hawthorn Butterfly.* The English name almost sufficiently describes this handsome species; it is about the size of the *Pontia brassicæ,* of a uniform white, with the nervures black: the under side is exactly similar. The caterpillar is black when young, but soon becomes partially clothed with hairs, and striped with reddish-brown on each side; it eats the leaves of the hawthorn, and will attack fruit-trees; the insect is by no means common, and is principally found in the south.

LEPIDOPTERA.   SPHINGIDÆ.

SMERINTHUS.

Generic Distinctions.—See page 135.

SMERINTHUS OCELLATUS. (Plate X.) *Eyed Hawk-Moth.* The male of this beautiful insect measures two inches and a half, and the female often exceeds this by an inch; the anterior wings, which are very acute at the tip, are grey, tinged with rose-colour, and variegated with brown; the posterior pair are rose-colour, with the anterior margin grey, and the hinder part tinged with that colour; it has also a large ocellus with a blue iris, and a large bluish-brown pupil placed in a patch of deep black; the head and thorax are nearly of the same colour as the surface of the upper wings. The caterpillar is green, the sides tinged with blue, and most of the segments marked with a white oblique stripe on the sides; the head is bordered with yellow; it feeds on the willow, poplar, apple, etc. The moth is common in the south.

DEILEPHILA.

Generic Distinctions.—Antennæ rather short, and thickening into a distinct club, with a terminal hook; *proboscis* long and spiral; *anterior wings* not very acute at the apex, the hinder margin slightly rounded.

DEILEPHILA EUPHORBIÆ. (Plate XII.) *Spotted Elephant Hawk-Moth.* The upper wings are olive-brown or greenish, with a rose-coloured irregular band, extending from near the base to the apex, and the margin of the same colour; the hinder wings are rose-coloured, with a white mark near the body, the base and a transverse band being black; the head and antennæ are white; the thorax olive-green; the body similar, with the sides of the three first segments white; the whole of the under side rose-colour. The caterpillar is black, with numerous yellowish spots in transverse lines; others red and cream-colour; the head, a line on the back, and the legs, are red; it feeds on the *Euphorbia Cyparissias,* cypress-leaved spurge. The moth is rare.

DEILEPHILA LINEATA is somewhat similar in colouring, but is rarely found.

———

LEPIDOPTERA.　SPHINGIDÆ.

MACROGLOSSA.

Generic Distinctions.—Antennæ clavate; *proboscis* of great length; *wings* opake; *body* having a tuft of scaly hairs at the extremity.

MACROGLOSSA STELLATARUM. (Plate XII.) *Humming-bird Hawk-Moth.* This, the only English specimen of the genus, measures nearly two inches; the anterior wings are dusky brown, with several transverse bands, the two middle ones being black, with a spot of the same between; hind wings rust-yellow; body of the same colour as the upper wings, variegated with tufts of black and white. The caterpillar is green, with numerous white points, and two white rays on the sides ; the horn is long, acute, and straight, yellowish at the tip, and blue at the base ; it is found on the *Galium*, or bedstraw. This curious and pretty insect is of frequent occurrence, and may be seen darting about in the day-time, from flower to flower, poising itself like a humming-bird over the blossoms, and extracting the juices with its long spiral proboscis ; while thus suspended, the vibration of the wings is so rapid as to occasion a humming noise.

LEPIDOPTERA. BOMBYCIDÆ.

LASIOCAMPA.

Generic Distinctions. — *Antennæ* bipectinated in the male, and merely serrated in the other sex; *mandibles* and *trunk* wanting; *palpi* short and hairy; *wings* entire; *body* tufted in the male.

LASIOCAMPA QUERCUS. *Oak Egger-Moth.* Colour deep chestnut-brown in the male, rather paler in the female ; with a yellowish band running across all the wings, and shading off towards the margin into the ground-colour ; on each side of the anterior wings is a small white spot. The caterpillar is yellowish, with greyish-brown hairs, and a white band on each side, also an interrupted row of white spots on the back ; it feeds on the oak, willow, etc. The moth is very abundant.

LEPIDOPTERA. ARCTIIDÆ.

HYPERACAMPA.

Generic Distinctions. — *Antennæ* simple; *palpi,* terminal joint shorter than the others ; *proboscis* long; *wings* very opake, and entire at the edges.

HYPERACAMPA DOMINULA. *Scarlet Tiger-Moth.* This beautiful insect has the primary wings black and shining, ornamented with about a dozen cream-coloured spots ; the secondary pair are of a fine crimson, with three irregular black spots ; the body is of a similar colour to the wings, but great variety occurs in the colouring of this pretty moth, the under wings being sometimes pale yellow, at others dull brown, without any vestige of the usually gay tints. The caterpillar is black, with three longitudinal yellow lines ; the black portion has many small blue tubercles with greyish hairs. The moth is not common.

CERURA.

Generic Distinctions. — *Antennæ* bipectinate; *palpi* four, and small; *proboscis* short, and nearly straight; *legs*, the anterior pair have an unusual appendage in the form of a lobe, long and compressed, attached to the inner side near the base.

CERURA VINULA. *Puss Moth.* The expansion of the wings, which are diaphanous (semi-transparent), is about three inches, the anterior pair being greyish-white, with many and varied dusky lines, and the base spotted with black ; the nervures are of a yellowish-brown colour, and strongly marked ; the hinder wings are ashy-brown, rather white at the edge, having an obscure crescent on the disc, and a few dark spots on the posterior margin. The caterpillar is remarkable for its grotesque form ; it has a large, singular-looking head, which it holds in an elevated position, and a forked tail, which is likewise kept in an erect posture. Isaac Walton's description, though curious, is tolerably accurate : — " His lips and mouth somewhat yellow ; his eyes as black as jet ; his forehead purple ; his feet and hinder parts green ; his tail forked and black ; the whole body stained with red spots, which run along the

neck and shoulder-blade, not unlike the form of a cross."
The cocoon made by this caterpillar is remarkable for its
strength, being composed of particles of wood united by a
kind of gum ; to enable the moth to pierce this case, it is
said to be furnished with a bag of acid, which dissolves the
gummy particles, and renders egress less difficult than
would be imagined from the nature of the obstacle ; it feeds
on the willow and poplar, and is not uncommon both in
England and Scotland.

LEPIDOPTERA. NOCTUIDÆ.

THYATIRA.

Generic Distinctions.—*Antennæ* simple; *maxillæ* as long as the
antennæ; *palpi* considerably apart, terminal joints short and equal,
middle one long and thick; *superior wings* moderately wide and
acute at the tip; *body* robust; *thorax* furnished with a crest.

THYATIRA BATIS. (Plate XIII.) *Peach-blossom Moth.*
The upper wings of this beautiful insect are brown, with
dark transverse lines, each having five rose-coloured spots ;
a large one at the base clouded with brown, two near the
tip, one on the posterior angle, having a brown spot in the
middle, and a small one near the hinder margin ; the inferior

wings are dusky, inclining to yellow, and having a pale line near the middle. The caterpillar is of a peculiar form, having a protuberance on the back, a little behind the head, cleft at the summit into two branches, and triangular elevations along the back; the colour is rust-brown; it feeds on the bramble. The moth is not very common, and is found near woods.

MELANIPPE.

Generic Distinctions.—*Antennæ* slender and simple; *palpi* short, slender, and acute, terminal joint small and pointed; *proboscis* rather long.

MELANIPPE HASTATA. *Mottled Beauty.* This species measures nearly an inch and a half; the ground-colour is white, with a broad black line or band round the hinder margin of all the wings, spotted and interrupted on the inner side; across the middle of the upper wings is also an irregular black band, nearly divided in the middle by the ground-colour; the base has another band interrupted by a white crescent; the under wings have a cross line of black angular spots. The caterpillar is dusky brown, with waving yellow lines on the sides. The moth is not very common.

LEPIDOPTERA. GEOMETRIDÆ.

OURAPTERYX.

Generic Distinctions.—Antennæ simple; *palpi* short; *proboscis* nearly as long as the antennæ; *wings,* anterior very acute, posterior much prolonged.

OURAPTERYX SAMBUCARIA. *Swallow-tail Moth.* The wings measure about two inches and a half; the colour delicate sulphur, shading into satiny-white towards the base; the surface marked with numerous streaks, placed transversely; two lines of deep yellow run across the anterior wings, and one across the under pair; at the base of the tail-like projection are two small blackish spots. The caterpillar is reddish-brown, with darker longitudinal lines; it feeds principally on the elder, willow, and lime-tree, and is common in many parts of England.

LEPIDOPTERA. TORTRICIDÆ.

HYLOPHILA.

Generic Distinctions.—Antennæ setaceous ; *anterior wings* very broad at the base, contracted in the middle, and again dilated at the hinder extremity.

HYLOPHILA PRASINANA. *Green Silver-lines.* The upper wings light green, with three oblique stripes of silvery white ; the hinder wings yellowish-white ; the margin of the upper wings is tinged with red.

HYLOPHILA QUERCANA. *Scarce Silver-lines.* The largest insect of this group, the wings sometimes extending to two inches ; the colour of the anterior wings deep grass-green, the latter traversed by two oblique white lines ; hinder wings glossy white ; palpi and antennæ tinged with red. The caterpillars of both these moths are light green, with a yellow line on each side, and two red streaks behind ; they feed on a variety of common trees. This moth is rare, but has been obtained in the neighbourhood of London.

LEPIDOPTERA. GEOMETRIDÆ.

ABRAXUS.

Generic Distinctions.—*Antennæ* moderately long and simple ; *palpi* short; *body* and *thorax* slender ; *wings* broad and rounded, edge entire.

ABRAXUS GROSSULARIATA.   (Plate XIV.)   *Gooseberry Moth.*   This insect measures nearly two inches ; the upper wings are white, with two bright yellow bands, and six transverse rows of black spots ; the under wings have a few scattered spots on the disc, and a series of larger size round the border ; the body is also yellow, spotted with black. The caterpillar is very similar in colour to the perfect insect, being white tinged with blue, the under side yellow, and the whole spotted with black.   The moth abounds both in England and Scotland, where the gooseberry and currant (the appropriate food of the caterpillar) are found.

HOMOPTERA. CICADIDÆ.

CICADA.

Generic Distinctions.—*Antennæ* six- or seven-jointed, very short and setaceous.

CICADA SPUMARIA is of a brown colour, with two white

spots on the upper wings. The larvæ of this insect is well known as discharging a kind of frothy matter, called by country-people "cuckoo-spit," in the midst of which they generally live, and find in it a protection against their enemies : perhaps, also, this moisture may defend them from the sultry beams of the sun. Some of the family possess the power of making a chirping sound, which has been celebrated by many poets; and in reference to its being produced only by the males, an old Rhodian bard says,—

> " Happy the Cicada lives,
> Since they all have voiceless wives."

DIPTERA. ŒSTRIDÆ.

ŒSTRUS.

Generic Distinctions.—Antennæ very short ; *proboscis* and *palpi* obsolete ; *thorax* smooth; *body* in the female very long.

ŒSTRUS EQUI. Head yellowish-white ; thorax yellow, with elevated hairs on a bluish point ; the end of the body is reddish, with two black spots ; the wings have a band in the middle and two small black points at the extremity.

These insects have the appearance of large Flies ; their

bodies are often hairy and ornamented with bands of various colours, the wings very strong, and the mouth nearly obsolete. The habits of these little creatures are very remarkable, each species being confined to its own peculiar quadruped, the horse, the ass, ox, etc. The larvæ reside either in the stomach, or beneath the skin of cattle, which seem to possess an instinctive dread of the presence of this insect, which is the more singular as it inflicts no pain, merely gluing its eggs to the hair. The larvæ when hatched burrow into the poor animal's back, gradually forming a protuberance of more than an inch in diameter, within which they live until attaining maturity. It is a curious fact that their spiracles, or breathing pores, are placed, not as usual on the sides, but at the extremity of the body, in order, as it would seem, to avoid the necessity of having a large orifice in the protuberance, which would interfere with the comfort of its temporary inhabitant. The *Œstrus* of the sheep is small, and of a greyish colour ; it places its eggs in the nostrils of that animal, whence the larvæ ascends into the head, and, when full-grown, falls down and assumes the pupa state on the ground.

Plate XV

N. Franke. del.

Auris Rosæ   2 Notonecta glauca   3 Culex pipiens   4 Cheironomus plumosus
5 Tabanus Bovinus

TABANUS.

Generic Distinctions.—Antennæ scarcely longer than the head, the last joint thick and crescent-shaped; *eyes* very large; *wings* horizontal, and triangular; *body* conical.

TABANUS BOVINUS. (Plate XV.) Head greyish-white, with the eyes of a shining green; the thorax and body blackish-brown, with spots.of a red tinge; the wings are transparent and veined with brown: it is a large and handsome insect.

The flight of some species in this genus is noiseless; their thick proboscis, armed with six lancets, inflicts considerable pain, and their pertinacity is so great, that it is difficult to drive them away; the horse lashes its tail, and tosses its head in vain, the *Tabanus* remains firm till its thirst is satiated. It is the female alone that possesses this bloodthirsty propensity; the males frequent flowers. The larva resides in the ground: it is long and cylindrical, narrow towards the head, which is small, and armed with two hooks.

MUSCA.

Generic Distinctions.—Antennæ rather short, the third joint much

N

longer than the two first, with a seta or bristle, often feathered; *palpi* filiform; *wings* distant.

MUSCA DOMESTICA. *Common Fly.* In this well-known species the antennæ are black, with a long terminal joint and bearded seta; the eyes brown; the fore part of the head white, and the remainder black; the thorax and body blackish above, and pale yellowish-brown underneath.

MUSCA VOMITORIA. *Bluebottle Fly.* Head yellowish, with brown eyes; the thorax black, and the thick short body of a deep brilliant blue, whence its common name; the young are deposited in the form of eggs, which are hatched in the course of two hours.

MUSCA CARNARIA. *Blow Fly.* Head yellow, the eyes reddish; antennæ plumose; the thorax is grey, and has four white lines; the body black, shining, and spotted with white. This last-mentioned species is very troublesome, from the constant endeavour to deposit its offspring on our animal food; but this propensity also renders it of great service, in consuming dead and decaying animal matter. In order to facilitate the discharge of this office, the Almighty Creator has endowed the females with the power of hatching their young in their own bodies, that they may be immediately ready to fulfil the important duty assigned them.

They deposit 20,000 young in the space of a few days; the young grubs increase in weight two hundred times in twenty-four hours, in consequence of their great voracity, so that they must assist wonderfully in clearing the earth of putrid substances; in five days they acquire their full size, and about the same time is spent in the pupa state, so that in a fortnight there are descendants of the first brood in existence. The larvæ of the *Muscidæ* are thick, fleshy, cylindrical, smaller towards the head and truncated at the other extremity; they are destitute of legs, moving by means of the hooks of the mouth.

DIPTERA. TIPULIDÆ.

CHIRONOMUS.

Generic Distinctions.—*Antennæ* beautifully feathered in the male; *body* slender and elongated; *wings* white, with three obscure points, from which arise pale transverse bands.

CHIRONOMUS PLUMOSUS, the *Midge*, is much smaller than the Gnat, which it so greatly resembles; they are often confounded together, but the former wants the long proboscis of the latter insect, and is perfectly harmless. The English name is equally applied to several genera, and in-

N 2

cludes above a hundred British species.   The larvæ reside
in water like those of the Gnat; that of *C. plumosus* is
found in stagnant water, and is called the Blood-worm,
from its red colour. Réaumur mentions that he found one
species enclosed in small paper-like cases of a brown colour;
they were spindle-shaped, composed of silk, and somewhat
resembled an oblong seed.   Midges may constantly be
seen in summer, hovering in moist situations, alternately
rising and falling, with a motion like that of the *Ephemera.*
Some species are remarkable for the beauty of their colouring, and form interesting objects for the microscope.

Messrs. Kirby and Spence give a very interesting description of the manner in which the perfect insect escapes from
the puparium; the native place of the pupa being water,
whereas the Midge is an inhabitant of the air.   For the
extrication of the imago, it is necessary that the pupa
should remain quietly suspended at the surface, and that
the thorax, in which the opening for its exit is to be made,
should be at least level with it; and this is precisely what
takes place. By a most singular and beautiful contrivance,
not only is the pupa, which is specifically heavier than the
water, enabled to suspend itself without motion at the surface, but its thorax, which is the heaviest end, is kept

uppermost. This is effected by the property which the centre of the thorax has, of repelling water ; hence, as soon as the pupa has once forced this part of the body above the surface, the water retreats from it on all sides, leaving an oval space in the disc which is quite dry ; and the attraction of the air to the dry part of the thorax is sufficient to overbalance the specific gravity of the pupa. If a drop of water be let fall on the dry portion, it instantly sinks to the bottom, but soon returning to the surface, it remains suspended as before. Just previous to the exclusion of the Fly, the thorax is seen to split in the middle ; the air enters and forms a brilliant stratum resembling quicksilver, between the body of the insect and the pupa-case, and the former, pushing forth its head and fore legs, like the Gnat, rests the latter on the water, and in a few seconds extricates itself wholly from the puparium.*

* British Cyclopædia, article *Chironomus*.

CHAPTER VII.

JULY.

THE gradually increasing heat of the season during the present month produces many insects new to the student ; and if he be interested in the pursuit, much knowledge may be acquired, many observations made, and an abundance of specimens added to the cabinet. In *Lepidoptera* July is particularly fruitful, as sunshine seems the appropriate element of these delicate creatures ; and the thicker clothing, if we may so term it, with the more robust forms of the nocturnal tribes, scarcely reconciles us to the idea of their still fragile beauty being exposed to the chilly dews of night in our northern climate. But we know that they are well adapted to their assigned place in the creation ; and to the imaginative mind there is something agreeable in the idea of being surrounded by sights, as well as sounds, of life and beauty, even when we are generally unconscious of their presence. Surely this thought adds its mite to the testimony of an over-watchful Providence, who peoples even darkness itself with brilliant colours, and elegant forms ; and animates the period of sleep, the emblem of death, with sounds which are but another name for melody, as if to reassure our waking moments with a consciousness of His presence, who "never slumbereth nor sleepeth."

The smaller species of Moth are so very numerous that but a very disproportionate number can be mentioned in this little work; a full account must be sought for elsewhere. There are supposed to be not less than 2,000 species, many of very minute size ; and it has been truly said, that the study of the minute *Lepidoptera* is yet in its infancy; the time of their flight, too, is an obstacle in the way of all but very zealous entomologists, though it must not be imagined that their appearance is entirely confined to the night or twilight: this is the more appropriate, but not the sole period of their activity, and many species may be seen flitting in the sunshine with the diurnal *Lepidoptera*, and other warmth-loving insects. Mr. Bird gives an account, in the " Entomological Magazine," of the plan he adopted for taking Moths, which may afford a hint to those who are desirous of studying this interesting and extensive group. He says: "My success in obtaining *Lepidoptera*, to which I am particularly attached, I owe to the use of a lamp: during the moonless nights of summer I sit with a Sinumbra lamp, and perhaps one or two smaller ones, placed on a table close to the window. The Moths speedily enter the room, if the weather be warm, and I have had a levée of more than a hundred between the hours of ten and twelve. I have, for experiment's sake, sat up till three o'clock, when the whole heaven was bright with the rising sun, and Moths of various kinds have never ceased arriving in succession till that time. In the spring and autumn I have been frequently very fortunate, though generally having my patience sufficiently tried. If at any time of the year a warm mist pervade the air, there is almost a certainty of success ; but should any one be induced by this account to try the lamp, he must make up his mind to experience more unfavourable evenings than favourable. There is, however, this advantage in my sedentary plan of mothing, that it can be combined with reading or writing; and the intervals between the arrivals need not be lost."

The extensive family *Ichneumonidæ* presents many very

interesting particulars to the student, which are highly deserving of attention. Referring to the order *Hymenoptera* for the characteristics of the family, we find that the insects comprised in it are provided with an instrument called an ovipositor, which is composed of two external filaments enclosing a slender piercer, with which the little creatures bore a hole in wood or other substances, to form a hole for their eggs. This instrument in some species is longer than the whole body; such is the case with the *Pimpla mani-festator*, one of the largest British species, which places her eggs in the holes already occupied by the young larvæ of the Wild Bee, and, as these are usually of some depth, it is necessary that this Ichneumon, which is destined to prevent their too rapid increase, should be able to reach the cell in which they are placed. The means by which the *Pimpla* ascertains the presence of the food destined for her young, seems to be her antennæ, which she introduces into the hole before depositing her eggs; but whether by the sense of smelling, hearing, or feeling, appears undecided. Those species that place their eggs in the bodies of caterpillars are provided only with a very short ovipositor for this purpose, as a more lengthened instrument would be superfluous. The *Ichneumonidæ* are all parasitic upon other insects, and through their means thousands of caterpillars are destroyed, which would otherwise do great injury. One species preys on the *Aphidæ*, or Plant-Lice, and thus assists in checking the increase of these troublesome little insects. This is the species best known, being often seen in the garden, particularly the male, hovering over the rose-trees, or creeping under the leaves ; it is of small size, with long legs ; the wings obscurely coloured with bands and spots of black and red. The female pierces the skin of the *Aphis*, and deposits an egg in the wound, taking great care not to place more than one in each insect ; when the grub is hatched, it feeds on the body of the poor victim ; and when full grown, it spins no cocoon, being adequately protected by the hardened skin of the dead *Aphis*. In a few days, the now winged insect forces itself from its prison, and flies away to enjoy its brief existence, and add another to the countless myriads of happy beings. Another species is mentioned by Mr. Kirby, as serviceable in keeping down the numbers of the little Midge which attacks wheat when in flower. Having placed a number of the grubs of this mischievous insect on paper, he introduced a female Ichneumon amongst them ; she immediately began to pace about, vibrating her antennæ very briskly ; then fixing on one of the larvæ, she inserted her ovipositor; the poor grub when pricked gave a violent wriggle, and this operation was repeated till all had received the egg which would prove so fatal a deposit. In these instances the larvæ of the Ichneumon are solitary, but in some species they are gregarious, the parent depositing several eggs in the same caterpillar: this is the case with the Ichneumon, *Microgastor glomeratus*, which preys on the larvæ of *Pontia rapæ*, or Small White Butterfly. These caterpillars, when full grown, creep up to the corner of window-frames, or similar places, to undergo their change; but frequently, previous to this transformation, a great number of small white grubs issue from them (causing, of course, the death of the caterpillar): the little grubs immediately enclose themselves in oval cocoons, attached round their victims, looking like a mass of yellow silk. Some times the larvæ remain within the caterpillar until it is in the chrysalis state, when they come forth in their perfect condition; this depends on the species of Ichneumon, as well as on the growth of the caterpillar. It is a wonderful fact that these little parasites never attack the vital parts of the larvæ (as this would destroy their supply of food), until they have nearly attained their full size, when they at last destroy the organs of life, and then piercing through

the skin they begin to construct their cocoons, which in
some species are found on the blades of grass, in others on
trees. The following notice, taken from that interesting
little monthly publication, the "Naturalist," shows that insects even in the pupa state do not escape the Ichneumon.
In reply to query, it is said, "Some years ago, while collecting in Botany Bay Wood, on Chat Moss, I was rather
startled on hearing a strange sound from among the dry
leaves scattered on the ground (especially as vipers are not
uncommon there).   On looking closely, I perceived it came
from an Ichneumon in the act of piercing a leaf. On
seizing it, I was delighted to have ocular proof that they
will attack pupæ; the leaf contained a pupa, which next
season produced *Acronycta rumicis.* I believe few entomologists have witnessed the above; it is the only instance
I have met with during several years' collecting." These
insects vary much in size, some being among our largest
Hymenopterous insects, and destined to check the too rapid
increase of the Hawk-Moths, and other large *Lepidoptera*,
others are so minute as to be seen with difficulty; some are
even small enough to be parasitic on the eggs of insects.
De Geer found a mass of sixty eggs of some Lepidopterous
insect, not one of which had escaped the Ichneumon.   We

have thus seen that these insects attack the larvæ of Hymenopterous, Dipterous, and Homopterous insects, in the case of Wild Bees, Midges, and Aphides; other species of these orders are similarly affected, as well as the *Coleoptera*, though the greater number are appropriated to the order *Lepidoptera*. But it does not end here, for these parasites are themselves liable to the attacks of smaller species belonging to the same family, and the instinct is remarkable, which guides these second parasites to the particular caterpillars already made the prey of the larger Ichneumons, which are to furnish their own offspring with support. These interesting, though diminutive insects, are divided by Latreille into groups known by the number of joints in the palpi, but these cannot be investigated in this rapid sketch, which is only intended to excite an interest, leading to research both in works of more pretension, and, better still, in the works of nature, which will amply repay the student. What an immense field is open to his survey may be imagined from the fact that in this family alone, including merely the European species, a work of three thousand pages has been published by Gravenhorst. The colour and form of many species are highly beautiful.

———

COLEOPTERA. BYRRHIDÆ.

ANTHRENUS.

Generic Distinctions.—*Antennæ* terminated by a three-jointed club; *body* of a round and depressed form, the surface adorned with undulating bands of coloured scales.

ANTHRENUS SCROPHULARIA scarcely exceeds two lines in length; the head is black with a small white spot; the antennæ are reddish at the base but black at the tip; the thorax is black, with the sides whitish, and the hinder part red; the elytra are also black, with bands of white; the under side is covered with white scales.   The larvæ feed on animal substances, and are sometimes very destructive in museums; the perfect insect frequents flowers.

COLEOPTERA. STAPHYLINIDÆ.

XANTHOLINUS.

Generic Distinctions.—*Antennæ* placed near each other at the base, and in general bent suddenly, geniculated, as it is termed; *legs* short and strong; *head* and *thorax* somewhat square; *body* linear.

XANTHOLINUS FULGIDUS.   The head and thorax of this

species are of a glossy black, punctured on each side ; the
elytra are deep red, punctured, and clothed with short
hairs ; the antennæ, mouth, and legs are light red.

COLEOPTERA. MELYRIDÆ.

MALACHIUS.

Generic Distinctions.—Antennæ, joints a little projecting on the
inner side ; *palpi* short and filiform ; *mandibles* notched at the joint ;
thorax wider than the head ; some species are armed with a spine at
the extremity.

MALACHIUS MARGINELLUS is of a brassy-green colour,
with the sides of the thorax and tips of the elytra red.

MALACHIUS ÆNEUS, shining green, with the margins of
the elytra bordered with red ; it is about a quarter of an
inch in length.

COLEOPTERA. CLERIDÆ.

CLERUS.

Generic Distinctions.—Antennæ, club three-jointed ; *palpi* terminated in a hatchet-shaped joint; *thorax* narrower than the elytra; *body* elongated.

CLERUS APIARIUS is of a blue colour, the elytra red, with blue bands; the larva of this insect is found in bee-hives, where it devours the larvæ of the Bee, and does much injury; it is, however, rare.

CLERUS or TRICHODES ALVEARIUS. This species is very similar to the preceding, but the extremities of the elytra are red. The larva resides in the nest of the Mason Bee, feeding on the grubs of that insect.

COLEOPTERA. CURCULIONIDÆ.

HYLOBIUS.

Generic Distinctions.—*Antennæ* of twelve joints; *elytra* oblong; *legs* moderately long; *body* furnished with wings.

HYLOBIUS ABIETIS, which varies from five to nine lines in length, is of a pitchy-black colour, with many yellow spots on the elytra. It abounds in the fir-plantations of Scotland, making great devastation, by destroying the bark; in 1824 many young firs were entirely destroyed on Lord Carlisle's estates in Scotland by this insect. The mischief was at first supposed to be the work of mice, so completely was the bark stripped from the trees.

STREPSIPTERA. STYLOPIDÆ.

STYLOPS.

Generic Distinctions. — *Antennæ,* outer branch flattened and three-jointed.

STYLOPS DALII. (Plate III.) Body of a deep velvet-black, and yellowish at the sides ; legs brownish ; the wings white and iridescent, about a line and a half in length. This singular tribe of parasitic insects has been found in the bodies of Wild Bees and Wasps, where they live in the larva state. They have been divided into four genera, distinguished by peculiarities in the antennæ. Rossi, an Italian entomologist, seems to have first noticed the insect, and discovered its habits ; and Mr. Kirby, having found another species, investigated the subject minutely, and established the order *Strepsiptera* for their reception. He observed a specimen in the larva state in the body of a Wild Bee, and having drawn it out, found a white fleshy grub, a quarter of an inch in length. On endeavouring to extract another, he

o

says : " How greatly was my astonishment increased when, after I had withdrawn it a little way, I saw its skin burst, and a head as black as ink, with large staring eyes, and antennæ consisting of two branches, break forth and move itself briskly from side to side ; it looked like a little imp of darkness."   Mr. Dale, who seems to have attentively watched these little insects, found several in *Andrena barbi-labris*, a species of Wild Bee, and caught one flying over a hedge : he says, " It looked milk-white on the wing, with a jet-black body, and totally unlike anything else : it flew with an undulating motion." There is much relative to the natural history of these singular creatures, which still remains unascertained or doubtful. There are several species of the genus *Stylops* in England ; and North America and the Mauritius have also furnished specimens of the order.

LEPIDOPTERA.   NYMPHALIDÆ.

MELITÆA.

Generic Distinctions.—See page 97.

MELITÆA SILENE, *Small Pearl-bordered Fritillary*, greatly resembles *M. Euphrosyne*, described in May, but it is much smaller ; the surface is similar, and the difference on the under side consists in the ground-colour of the secondary wings being rust-brown, with the transverse band at the base and middle, not so light a yellow; in having three silvery spots in the central band, and five others, three of which are placed in a line on the anterior border, and the other two near the inner edge, and in having the ocellus towards the base black, with a red pupil. The caterpillar is black and spiny, one half of the spines being yellow, and the sides of the body having a light stripe. The Butterfly is not uncommon.

———

LEPIDOPTERA. NYMPHALIDÆ.

ARGYNNIS.

Generic Distinctions.—See page 159.

ARGYNNIS PAPHIA. (Plate VII.) *Silver-washed Fritillary.* The surface of this insect is very similar to that of the species *Adippe* and *Aglaia*, mentioned in June, the upper surface being of a bright yellowish-brown, variously streaked and spotted with black; on the under side the primary wings are paler, many of the black spots indistinctly seen, and the tips slightly tinged with green; the secondary wings are green, with a brassy lustre, and ornamented with stripes or bands of silver, not spots as in the other species. The caterpillar is light brown, with two dark lines on the sides, and long hairy spines : it feeds on the raspberry and violet. The insect is not uncommon.

LEPIDOPTERA. LYCÆNIDÆ.

POLYOMMATUS.

Generic Distinctions.—See page 123.

POLYOMMATUS ARGUS. *Silver-studded Blue.* Male, deep blue above, inclining to lilac, with a broad band of black round the hinder margin of all the wings; the nervures are also dark; the under side is bluish-grey, becoming deeper at the base, and adorned with many ocellated spots; on the hinder margin of the posterior wings there is a tawny orange band, containing six bright silvery blue spots crowned with a series of black crescents. The female is entirely brown above, with a tawny marginal band. The caterpillar is dull green, with a rusty line on the back, and oblique marks of the same colour, edged with white, on the sides; it feeds on broom, trefoil, etc. The butterfly is common in the south.

POLYOMMATUS SALMACIS. *Durham Argus.* This species is silky-brown above, the anterior wings having a white spot, and all of them an orange band, indistinct on the upper wings; the under side is greyish-brown, the anterior pair having a white spot, and a curved band of the same; these are succeeded by a band of orange spots. The hinder wings have a similar marginal band, and several scattered white spots, a large one near the centre, and a series behind, connected with the yellow band by a broad central patch: most of these spots have a minute dusky pupil. This insect is principally found in Durham.

POLYOMMATUS ARTAXERXES, which appears principally to differ from the last in being of a much darker hue, and the spots having no pupils, is principally met with on Arthur's Seat, near Edinburgh: from this circumstance it is called the Scotch Argus; but very little seems to be ascertained respecting the history of the insect in its immature state, though the butterfly is in great request, from the extreme rarity of its appearance in any other locality.

POLYOMMATUS CORYDON. *Chalk-hill Blue.* This is one of the larger species, measuring an inch and a half. The surface of the male is a very light silvery blue, with a fine silky lustre; the hinder margin of all the wings having a blackish band, and the inferior pair a series of dusky spots. The surface in the female is brown, each wing having a pale central spot; the hinder wings in both sexes whitish, the ocellated spots being arranged in two curved bands in the centre; between these is an angular white mark, and on the hinder margin a series of black spots, with a white iris, half surrounded by a streak of orange: an oblong patch of white connects this series with the central band.

LEPIDOPTERA. NYMPHALIDÆ.

' VANESSA.

Generic Distinctions.—See page 129.

VANESSA POLYCHLORUS. *Great Tortoise-shell.* This species bears much resemblance to *V. urticæ,* described last month, but is much larger, sometimes exceeding two inches and a half; the upper side is dark orange; the anterior wings have two large quadrate (four-sided) spots of black, with two smaller ones near the base, the same number of rounded spots on the disc, and another near the hinder angle. The inferior wings have one large black spot near the middle of the anterior margin; both pair have a deep black border, ornamented with crescents which are blue on the hinder wings, and bounded with two lines of pale yellow; on the under side, half the surface is dark brown, the remainder grey, marked with waving lines of brown, and a faint series of bluish crescents towards the tip: there are one or two pale spots on the wings. The caterpillar is bluish-brown, with a lateral stripe of orange; the spines are slightly yellow: whilst young, the larvæ live together in a silken web, but disperse after they have changed their first skin ; they feed on the willow and elm. This is not one of the most common Butterflies, though occasionally found in great abundance in the southern counties and in the Isle of Wight.

LEPIDOPTERA. NYMPHALIDÆ.

HIPPARCHIA.

Generic Distinctions.—See page 98.

HIPPARCHIA SEMELE. *The Greyling.* This is one of the largest species of the genus, sometimes reaching to two inches and a half across the wings ; the greater part of the surface is brown, varying much in depth of colour; the female has a wide band of pale yellow near the hinder edge of the anterior wings, in which are placed two ocelli. The male has only a yellowish patch round each ocellus ; the base of the hinder wings, and a posterior border, are brown in both sexes, the intermediate portion being pale or reddish-yellow, with a small ocellus on the under side, the anterior wings are tawny at the base ; the disc pale yellow, with two ocelli, and the margin brownish ; the posterior wings are clouded with white, and dark brown ; the caterpillar is light green.   This insect frequents stony and rocky places, occurring plentifully in such situations both in England and Scotland.

HIPPARCHIA HYPERANTHUS. *Ringlet Butterfly.*   The whole upper surface is of a dark brown colour, generally having two or three eye-like spots on each wing; the under side is pale brown, with large ocelli near the margin ; sometimes these are nearly obliterated, but usually they are very conspicuous.   The caterpillar is greyish-white, with a brown line on the back : it feeds on the *Poa annua*, or meadow-grass.   The butterfly is found pretty abundantly in the open part of woods, meadows, and corn-fields.

HIPPARCHIA POLYDAMA. *Marsh Ringlet.*   This species measures about an inch and a half across the wings ; the surface is rusty yellow, tinged with brown ; the anterior wings having one or two ocelli towards the hinder margin ; the inferior wings are greyish-white round the outer margin, with an ocellus near the body, and sometimes one or two others on a line with it; underneath, the primary wings are greenish-brown at the base, bright yellowish-brown in the middle, and grey at the apex, with a small white bar and a few ocelli; the base of the hinder wings is brown, beyond which is a very irregular white band, the space beyond being greyish-brown, with five or six ocelli. The caterpillar is dark green, with a darker line on the back. The perfect insect is found on marshy heaths.

HIPPARCHIA DAVUS very closely resembles the preceding. It is extremely rare in England, but common in many parts of Scotland.

HIPPARCHIA CASSIOPE, *Mountain Ringlet*, measures about sixteen lines; the colour is dark brown, with a silky gloss; the upper wings having a red band near the apex, marked with black spots; the hinder pair have a short band of continuous red marks, each bearing a small black spot; the under side of the anterior wings differs from the upper, only in being tinged with rust-red; the hinder wings are ash-brown, having three black spots with a reddish iris. This insect is principally found in the mountains of Cumberland and Westmoreland, also in the hilly districts of Scotland. There are two other species rarely found excepting in Scotland, the *H. blandina* and *H. ligea.* The former is taken in great profusion in August, near Brodick, Isle of Arran.

————

LEPIDOPTERA. LYCÆNIDÆ.

THECLA.

Generic Distinctions.—See page 128.

THECLA QUERCUS. (Plate IX.) *Purple Hair-streak.* Size about fourteen lines; colour of the upper surface dark brown, faintly glossed with purple in one sex; the other with a large patch of deep glossy purple at the base of the upper wings; on the under side the wings are ash-grey colour, with a silky lustre, and traversed by an undulating white streak, beyond which is a double series of whitish crescents, and a few dusky spots on the primary wings; the secondary pair are ornamented with two reddish spots, one having an ocellus. The caterpillar is greyish-brown, with a brown head, and a row of yellow dots on the back; it is very common in oak-woods in the south.

THECLA W-ALBUM. *White-letter Hair-streak.* Upper side dark brown, with a silky gloss, the male having a greyish spot near the middle; the under side is light brown, with a small transverse white line near the hinder margin of the primary wings, and another towards the middle of the secondary pair, forming two acute angles resembling the letter W; behind this is a band of orange-red; the margin and the projecting tailed point are black, sometimes tipped with white. The caterpillar is green, with three red spots on each side of the lower segments, and a double series of small dots on the back; it feeds on the elm and blackthorn. The perfect insect is in general rare, but in some seasons is found in great profusion.

THECLA PRUNI. *Black Hair-streak.* This species has often been confounded with the former, which is more common; the difference on the upper surface seems to consist in the addition of three or four crescent-shaped red marks near the hinder margin; the colour of the under side is yellowish-brown, marked with an irregular silvery line near the hinder margin of the primary wings, and across the middle of the secondary pair; behind this is a row of black spots, and on the margin a series of black crescents; the upper wings have the spots red. The caterpillar is green, with longitudinal whitish rays, and numerous short transverse lines; the head is small and yellow.

APATURA.

Generic Distinctions.—*Antennæ* long, with an egg-shaped club; *palpi* long, and projecting beyond the head, where they meet and form a kind of beak; the basal and terminal joints of nearly equal length, the intermediate one very long; *wings* somewhat triangular, edge of the primary pair nearly entire, the other slightly scalloped.

APATURA IRIS. (Plate VIII.) *Purple Emperor.* Surface of the wings dark brown, changing, as the light shines on it, to a very rich purple: this brilliant colour is not seen in the female, whose wings are of a paler brown. In both sexes there are four patches of white on the upper wings, interrupted by the nervures; the hinder wings are traversed by a band of white, a tawny streak running round them near the margin, and the inferior pair have a black spot surrounded by a red ring; on the under side the superior wings are rust-red, with a large ocellus near the hinder angle, and two black spots near the base; the hinder portions of the under wings are greyish, with a faint undulating brown line near the margin; all the wings have the white marks similar to those of the surface. The caterpillar is pale green, with a yellow stripe down

Plate VIII

the outer side of two long horns, into which the head is divided behind; it feeds on the oak. This fine insect is much prized, both for its beauty and the difficulty with which it is obtained; it is found in the southern counties, but as it fixes its throne on the summit of a lofty oak, from whence it takes very high flights, there are many difficulties to overcome in the capture; and a pole, twenty or thirty feet in length, with the bag-net attached, is necessary for the purpose. The wings of these insects are stronger than those of any other British species; thus fitting them for the powerful flight which their habits require.

———

LEPIDOPTERA. NYMPHALIDÆ.

LIMENITIS.

Generic Distinctions.—Antennæ thickening gradually from the middle, almost to the apex, the club being long and slender; *palpi,* basal joint the shortest and nearly oval, the second very long, and the terminal one egg-shaped, and ending in a point; *wings* rounded and entire.

LIMENITIS CAMILLA. (Plate VIII.) *White Admiral.* This pretty insect measures about two inches; the colour is dull black above, marked with dark spots; both wings traversed by a broad white band, divided in the middle of the upper pair, which have also several white spots; on the hinder wings are two rows of obscure dark spots, between the band and the hinder extremity. The prevailing colour on the under side is brownish-yellow, all the white spots of the surface being visible, with the addition of a few others; the base of the wings is tinged with blue. The caterpillar is green, with obtuse fleshy projections on the back, of a reddish colour, and fringed with hair; it feeds on the honeysuckle. The perfect insect is one of our rarer species; it is found in the glades of woods, and is noted for its graceful flight. Many specimens were taken this summer in Burnham Beeches, Buckinghamshire.

LEPIDOPTERA. HESPERIDÆ.

PAMPHILA.

Generic Distinctions.—See page 131.

PAMPHILA LINEA. (Plate X.) *Small Skipper.* This insect is of small size; the surface of the wings fulvous and without spots; the hinder margin and nervures black;

1. Thymale alveolus   2. Pamphila linea   3. Anthrocera Filipendulæ.
4 Smerinthus ocellatus   5. Smerinthus Tiliæ .

the under side of the primary wings paler than the surface, shading into grey at the tip, and brown at the base; the secondary wings are ash-grey, with a fulvous patch; the male has in addition a black oblique line on the disc of each anterior wing. The caterpillar is deep green, with a dark line on the back; it feeds on some species of grass. The perfect insect is of frequent occurrence; there are two other species found occasionally, *P. coma* and *P. Actæon*.

LEPIDOPTERA. ANTHROCERIDÆ.

ANTHROCERA.

Generic Distinctions. — *Antennæ* simple, slender, and of great length, thickening into a curved club; *palpi* rather long, terminating in a point, and thickly clothed with hair.

ANTHROCERA FILIPENDULÆ (Plate X.), *Six-spotted Burnet Moth*, measures about an inch and a half; the upper wings are greenish-black, with six red spots on each; the under pair carmine-red on both sides; the hinder margin with a black border. The caterpillar is yellow, with three rows of black spots on the head and others on the sides; it feeds on grass and other common plants.

ANTHROCERA LOTI. *Five-spotted Burnet Moth.* Very

similar to the preceding, but smaller, and with only five spots on the upper wings. The caterpillar is green, with two longitudinal bands on each side, and a yellow dot on each segment; it feeds on the honeysuckle and other plants. The moth is less common than the preceding.

There are several species of this pretty genus, all noted for their brilliant colours, generally bluish-black and bright red, the latter hue predominating on the lower wings and forming spots on the upper. They are occasioually very numerous: I have found them in profusion near Ventnor, Isle of Wight, and they are also found in Scotland.

LEPIDOPTERA. SPHINGIDÆ.

SMERINTHUS.

Generic Distinctions.—See page 135.

SMERINTHUS POPULI. *Poplar Hawk-Moth.* This species differs from all those mentioned before, in having the outer margin of all the wings dentated; the colour is greyish-brown, occasionally inclining to rusty-red and grey, with bands and rays of a deeper hue than the ground-colour; the upper wings have a white mark in the centre; at the base of the inferior pair is a rusty patch, and sometimes a white spot near the middle. The caterpillar is green, with oblique yellow or white stripes, the head bordered with yellow, and the horn at the extremity of the body is of the same hue; it feeds on the poplar. This is the most common of the *Sphingidæ,* both in England and Scotland.

LEPIDOPTERA. HEPIALIDÆ.

ZEUZERA.

Generic Distinctions. — *Antennæ* setaceous, short, pectinated in the males, simple in the females; *palpi,* obsolete; *body* long.

ZEUZERA ÆSCULI. *Wood Leopard-Moth.* This beautiful species is snowy-white; the wings but sparingly clothed with scales; and the nervures yellowish; the whole surface thickly sprinkled with dark-blue spots; the hinder wings are white, and faintly spotted at the base, with a distinct series of spots round the margin; the male measures about two inches, the female is often a third larger. The caterpillar is light yellow, with a double series of black spots across each segment; it is not very common.

P

ENDROMIS.

Generic Distinctions.—*Antennæ* in both sexes bipectinated; *head* clothed with long hairs; *wings* entire, large, and rather transparent.

ENDROMIS VERSICOLOR. *Kentish Glory.* The upper wings in this species are rusty-red, inclining in some parts to grey, each with two transverse bands of black, edged on one side with white; there are also other white marks, and a small black crescent near the centre; the inferior wings are tawny-yellow, with a waved dusky line in the middle, a small crescent, and a few white spots. The caterpillar is pale green; the under side spotted with black, and the sides marked with oblique stripes of pale yellow; the form is peculiar, bearing some resemblance to the *Sphingidæ*, having an elevation where they have a horn; it feeds on a variety of trees, and must be ranked with the rarer species, being found principally in the woods of the south of England.

LEPIDOPTERA. NOCTUIDÆ.

THYATIRA.

Generic Distinctions.— See page 170.

THYATIRA DERASA. *Buff Arches.* The prevailing colour in this species is light yellowish-brown; the upper wings having two oblique white bands, the space between which is clouded with brown and white; there is also a transverse series of fine zigzag lines of a brown hue, forming acute angles on a whitish ground; the hinder margin is brown, with two rows of small white arches, surmounted by a white line; the hinder wings are dusky, with a tinge of ochre. The caterpillar is yellowish-green, with brown spots and lines; it seems to be a general feeder. The moth is found pretty frequently near woods and shady lanes.

LEPIDOPTERA. NOCTUIDÆ.

CATOCALA.

Generic Distinctions.—*Antennæ* long and setaceous; *proboscis* as long as the antennæ; *palpi*, middle joint much longer than the others, and densely clothed with scales; *body* thin at the extremity; *thorax* slightly crested.

P 2

CATOCALA FRAXINI.   *Clifden Nonpareil.* This very
handsome species sometimes measures four inches; the
upper wings are light grey, marked with undulating lines of
brown; the under wings are brownish-black, with a broad
curved band of light blue across the middle; they are also
deeply indented.   The caterpillar is ash-coloured, rather
inclining to yellow, and sprinkled with minute black dots;
the head is green, and the eighth segment has a protuberance of a bluish-black colour; behind this is an oblique
black line; it lives on the ash, oak, elm, poplar, etc.   This
species is not so common as the following, and a good
specimen is a valuable acquisition.

CATOCALA NUPTA.   (Plate XIV.) *Red Underwing.*
Upper wings dark grey, with transverse waved streaks and
spots of brown and yellow; the hinder border has a series
of crescent-shaped spots, and two waved dusky lines; the
under wings are of a deep carmine-red, with a curved black
band near the middle, and a broad marginal band of the
same: the under side of the superior wings is black, with
three transverse bands of white, the outer one forming an
acute angle near the inner margin. The caterpillar is narrow
at both ends; the colour grey, marked with brown, greenish
underneath, spotted with black, and a row of small tubercles

Plate XIV

1 Plusia chrysitis   2. Catocala nupta   3 Abraxas Grossulariata   4 Pterophorus pentadactlus
5 Alucita hexadactyla

on the back; it feeds on the poplar and willow. The moth is frequent in the southern counties.

LEPIDOPTERA. GEOMETRIDÆ.

HIPPARCHUS.

Generic Distinctions.—*Antennæ* pectinated in the male, simple in the female; *proboscis* spiral, and shorter than the antennæ; *palpi* meeting at the tip, basal joint very short, second long, the last lanceolate; *wings*, the lower pair are covered by the upper when at rest, the margin of the former dentated.

HIPPARCHUS PAPILIONARIUS. *Emerald Moth.* This species measures more than two inches; it is of a deep grass-green, with two rows of whitish spots across both wings; on the disc of each is an obscure crescent-shaped spot of a deeper green than the rest. The caterpillar is green, with a yellow line on each side, and reddish warts on the back; it feeds on the elm, lime, and beech. The moth is not very common, either in England or Scotland.

PIOPHILA.

Generic Distinctions.—*Antennæ* small, with a single bristle; *head* roundish, without a frontal projection; *legs* moderately long and slender.

Piophila casei. *Cheese Fly.* This insect in its larva state is a small fleshy grub, of an elongated form and white colour, found in decaying cheese. Epicures do not hesitate to eat these little creatures, thinking that they are bred spontaneously from the best parts of the cheese, whereas they are produced from eggs laid by the parent Fly, in a similar manner to the Blow Fly, from whose offspring on their meat these lovers of good eating would turn away with disgust. The Fly is about the size of the common House Fly, of a shining, blackish-green colour, with transparent wings; the legs are varied with black and ochre; the female is provided with an ovipositor, with which she pierces the cheese, and deposits an egg in the hole; these are soon hatched, and feed on the cheese during the larva state. When they have acquired their full size, they leave their food and undergo the transformation into a chrysalis; and after remaining in this state for some time, the enclosed Fly breaks that part of the outer covering which defends

the head into two parts, at the same time throwing off a
thin membrane which covered the body. At first the
wings are scarcely visible ; the insect, however, runs about
very quickly, and by degrees the wings assume their full
size. The leaping powers of the larvæ are very wonderful,
the leap being performed in the same manner as that of a
salmon, by taking hold of the tail with its mouth, con-
tracting the rings of the body, and then letting go the tail.
If a viper had equal powers in proportion to its size, it
would spring nearly a hundred feet.

CHAPTER VIII.

AUGUST.

THE young student has, I hope, gained sufficient information, and experienced sufficient pleasure from the study and research of the preceding months, to continue the pursuit with eagerness during the remainder of the season. He is now supposed capable of placing most of the insects he may find, in their proper order; in some, such as the diurnal *Lepidoptera*, he will be able to descend as far as the species, and thus to name the beautiful specimen correctly, unless it is a very rare insect; many moths will also be familiar to him, and the order *Coleoptera* understood with regard to its sectional and family peculiarities; some insight will have been gained into the habits of all classes of the insect world, and the student will not be liable to make the curious mistakes, or give way to the silly prejudices, of those who have never studied this interesting subject. Many persons have no idea that nearly all the insects they see have enjoyed three distinct modes of existence; that they all fulfil some important function in nature; and that the loss of one among the many different groups might cause infinite distress and confusion, by allowing the undue increase of others on which they depend for food; or, by not clearing the earth of corrupt animal and vegetable matter, cause disease to attack those who proudly or ignorantly demand the use of this apparently insignificant portion of the creation. The more attentively the subject is examined, the more plainly will it be seen that all are governed by wise laws; that not even the meanest insect is without its appointed work; well would it be for those who are at the summit of earth's created beings, would they endeavour to perform their share in nature's plan with as much fidelity as the little Burying Beetle, or the still more diminutive Ichneumon. "Order is Heaven's first law," and man alone uses the intellect which raises him above the creatures of instinct to rebel against it. This month is replete with beauty of every kind; fruit, flowers, and the insect world, all add to the charms of the season; and though a slight tinge of decay, mingled with the tints of the verdure, already reminds us of the approaching autumn, still much is to be enjoyed before the lover of nature must exchange the evening ramble for the fireside, and the book of nature for those which will assist him in studying it more accurately another year. This month produces one of the largest and most beautiful of our diurnal *Lepidoptera*, the lovely Camberwell Beauty, which is highly prized from the circumstance of its appearance being periodical, the cause of which has not been fully ascertained. " Until four or five years since, it had not been seen for forty years, and was then very abundant in many parts of the kingdom; in 1819, a few were taken in Suffolk, since which period it has not been seen."* Other species, both Moths and Butterflies, appearing at this time, are remarkable for their beauty and rarity; these will be mentioned in their proper places, and happy may the young entomologist consider himself who can add a specimen to his collection. Those who are poetically inclined will be induced, on seeing them, to quote the lines of Mrs. Barbauld :—

> " Lo ! the bright train their radiant wings unfold,
> With silver fringed, and freckled o'er with gold ;

* British Entomology.

On the gay bosom of some fragrant flower,
They, idly fluttering, live their little hour;
Their life all pleasure and their task all play,
All spring their age, and sunshine all their day."

A less agreeable insect begins to annoy us by its presence about the end of this month, belonging to the genus *Stomoxys* and the order *Diptera*. The species are small Flies, which frequent the windows of our rooms late in the summer, particularly when the weather is damp; they greatly resemble the common House-Fly, but are broader in form, and the proboscis is capable of giving a sharp wound, from which circumstance it has been commonly supposed that the Domestic Fly stings in the autumn; but the species is totally different. The specific names sufficiently testify their teasing propensity, *stimulans*, *pungens*, and *irritans* being members of this genus; the *Stomoxys calcitrans* is the most common.

The Tortoise Beetles, or *Cassidæ*, are an interesting and peculiar family, both from their form and habits; there are about twenty species in England, but of a small size compared with those of Brazil, which are often armed with two erect spines rising on the centre of the elytra, meeting together and forming an acute horn nearly half an inch in length. Some of the species are highly metallic, and though they lose their beautiful colours when dead, they may be restored by dipping the insect into hot water. The English name almost sufficiently describes the curious form of these Beetles, which greatly resembles that of a shield ; and as, when disturbed, they contract the antennæ and legs under the broad sides of the thorax and elytra (the head, too, being often quite concealed), they have the appearance of small Tortoises. The variety of form as well as colour in the Coleopterous tribes will strike every eye on first seeing a large collection of these insects, and nothing can well be imagined more frightful than some of them would appear, were their dimensions equal to that of many quadrupeds; as it is, they surprise and gratify, without exciting any feeling allied to fear. This idea is well expressed in the following lines :—

> " Their shape would make them, had they bulk and size,
> More hideous foes than fancy can devise ;
> With helmet heads and dragon scales adorn'd,
> The mighty myriads now securely scorn'd
> Would mock the majesty of man's high birth,
> Despise his bulwarks and unpeople earth."

COLEOPTERA. MORDELLIDÆ.

RIPIPHORUS.

Generic Distinctions.—*Antennæ* deeply pectinated in the male, those of the female serrated; *body* arched; *thorax* semicircular; *elytra* narrow at the tips, not meeting in a straight line, nor covering the wings.

RIPIPHORUS PARADOXUS, sometimes called *Mordella paradoxus*, resides in the nest of the Wasp, *Vespa rufa.* Mr. Kirby, in his Bridgewater Treatise, says, "Connected with the subject of parasites is a singular history communicated to me by Mr. Hope. In the month of August, 1824, he found more than fifty specimens of a singular little Beetle in a Wasps' nest; from their being found in cells closed by a kind of operculum, he conjectures that they lay their eggs in the grub of the Wasp, upon which they doubtless feed; but on opening some of the cells, he was surprised to find, instead of Beetles, several specimens of an Ichneumon; upon further examination he discovered that these last insects had been pierced in their turn, whilst in the chrysalis state, by a more minute species, of which he found more than twenty specimens flying about in search of their prey. From the above facts, Mr. Hope remarks, we have a convincing proof of a superintending Power, which ordains checks and counter-checks to remedy the fruitfulness of the insect world ; first the Wasp, a great destroyer of Flies and other insects, is prevented from becoming too numerous by the Wasp-Beetle ; then, lest it should reduce their numbers so as to interfere with their usefulness, this last is kept in check by the *Anomalon*, which in its turn, that it may obey the law, 'thus far shalt thou come and no further,' becomes the prey of another devourer." The *Mordellidæ* are all of small size, and occasionally variegated in colour.

LEPIDOPTERA. NYMPHALIDÆ.

VANESSA.

Generic Distinctions.—See page 129.

VANESSA ANTIOPA. *Camberwell Beauty.* This species sometimes measures three inches ; the upper side of both wings is a deep purplish-brown, having the appearance of velvet, bounded externally by a broad band of velvet black, in which is a series of large violet-blue spots ; beyond this there is a cream-coloured border, slightly waved on the inner side, and sprinkled with minute black spots ; the anterior border has two cream-coloured spots, and is mottled with yellow towards the base; the under side is shining, dark-brown, with waved lines of deep black, a small yellow spot near the middle of each wing, and two others on the border. The caterpillar is gregarious, and lives on various trees, such as the birch and willow: it is black, with a series of spots on the back, and some of the legs red.

LEPIDOPTERA. NYMPHALIDÆ.

CYNTHIA.

Generic Distinctions.—Antennæ with a short and abrupt club; *palpi* long; *wings* scarcely angular, the hinder pair rounded and simply scalloped, without any projection.

CYNTHIA CARDUI. (Plate VIII.) *Painted Lady.* Upper wings tawny-brown at the base, ochre-red in the middle, with a very irregular patch of black; a large portion of the apex is also black, adorned with five white spots, of which the inner one is the largest and placed obliquely; near the margin is a series of white crescents and a row of faint-yellow spots; the secondary wings are the same colour, with three rows of black spots behind, the first being composed of five round spots, the second of crescents, and the third of large patches placed on the projecting points ; the inner angle has a large black spot, with a streak of blue behind ; on the under side the primary wings are whitish at the base, and have a large white spot in addition to those corresponding with the upper side ; the tip is light brown, and the whole disc tinged with carmine intermixed with ochre ; the hinder wings are delicately variegated with light brown, greyish white and yellow; there is a series of ocelli towards the hinder extremity, and a row of purplish-blue crescents. The cater-pillar is very spiny, of a brownish-grey colour, with yellow lines on the sides ; it feeds on the thistle, nettle, and mallow. This beautiful insect is of large size; the under side deserves particular notice for its delicate colour and mark-ings ; it is not generally common, though sometimes appear-ing in considerable numbers ; it is found also in Scotland.

LEPIDOPTERA. LYCÆNIDÆ.

THECLA

Generic Distinctions.—See page 128.

THECLA BETULÆ. *Brown Hair-streak.* This is the largest species of the genus, sometimes measuring eighteen lines ; the upper side is a glossy brown ; towards the middle of the anterior wings, is a blackish mark, with a faint-yellow cloud beyond it, which is more clearly defined in the female, and of a deeper colour ; the secondary wings are of a similar hue to that of the superior pair, the projecting lobes being marked with reddish yellow ; the under side is tawny-yellow, inclining to red in some parts, with two small un-dulating white lines edged with black. The caterpillar is green, with yellow stripes on the back, and transverse rays on the sides, of the same colour ; it feeds on the birch, blackthorn, etc. The butterfly is not common.

LEPIDOPTERA. NYMPHALIDÆ.

ARGYNNIS.

Generic Distinctions.—See page 159.

ARGYNNIS LATHONIA. *Queen of Spain Fritillary.* Surface yellowish-brown, with numerous black spots of a rounded form beneath ; the primary wings are paler than above, but marked in a similar manner, and having a few silvery spots towards the tip ; the secondary wings are ornamented with above twenty silvery spots unequal in size ; seven of them of a semicircular shape form a row near.

Q

the hinder margin, before which is a transverse series of ocelli of a brownish colour with a silver pupil. The caterpillar is said to be greyish-brown, with a white line along the back : it is spiny ; the heartsease and some other plants form its food. The butterfly is small and rare.

LEPIDOPTERA. PAPILIONIDÆ.

COLIAS.

Generic Distinctions.—*Antennæ* short and robust, thickening into an obtuse club; *palpi* projecting beyond the head; *wings,* primary pair triangular, secondary rounded.

COLIAS EDUSA. *Clouded Yellow Butterfly.* The male is pale orange-yellow ; the upper wings with a wide black border, waved on the inner edge, and a rounded spot of the same in the middle of each ; the hinder wings are also margined with black, the ground colour is slightly mixed with green, and each has a spot of deep yellow ; on the under side, the upper wings are pale-tawny, greenish at the extremity, with a central black spot, and a faint series of blackish spots parallel with the outer edge ; the under wings are greenish, with a row of rust-coloured spots on the margin. The female has a few yellow spots on the black band of the upper wings ; examples of this sex are found, in which the parts generally yellow are greenish-white. The caterpillar is deep green, with a white line on each side, marked with blue and yellow dots; it occurs pretty frequently in the south, seeming to prefer the vicinity of the sea.

COLIAS HYALE. *Pale Clouded Yellow*. Rather larger than the preceding; the male of a fine sulphur-yellow; the female white, faintly tinged with sulphur; the upper wings are greyish at the base, marked with a black spot, and having a broad black band nearly divided by light spots; the under wings have a large round mark on the disc, and a few dusky spots near the outer edge; beneath, the upper wings are yellowish-white, orange at the tip, a black spot with a yellow centre on the disc, and a row of small dusky marks near the outer margin; the under wings are dull yellow with two silvery spots in the centre, surrounded with rust-red, and a curved row of small black marks. The cater-. pillar is green, with two white lines on the sides, and each segment marked with two series of black dots. The species is much more uncommon than the preceding, being chiefly found in Kent, Sussex, and Suffolk, near the coast.

There is one other species, which seems however a doubtful native, and differs principally from the last in having the black border extended to the hinder wings, that of the upper pair being entire ; and, though similar in this respect to the *C. edusa,* it is sufficiently distinguished by its paler colour. Further particulars are unnecessary, from the extreme rarity of the insect.

———

LEPIDOPTERA. LYCÆNIDÆ.

LYCÆNA.

Generic Distinctions.—See page 130.

LYCÆNA DISPAR. *Large Copper.* The male of this beautiful insect is bright copper-colour above, with a black margin round all the wings ; the base of the primary pair is also blackish, on the disc are two small black spots, and near the middle of the secondary wings is a curved black streak. The female has the upper wings broadly margined with black, two or three black spots on the disc, and a series of the same a little beyond the middle; the hinder wings are almost entirely black, except the nervures and a broad band near the apex, which are copper-coloured. Beneath, both sexes are similar ; the upper wings coppery, with spots similar to those on the surface of the female, but surrounded with a yellow ring; near the margin is another row of black spots, beyond which the colour is grey ; the under wings are ash-colour, tinged with blue, and having a coppery band on the hinder margin with a series of black spots on each side ; before this, is another row of black spots surrounded with bluish-white, then a transverse black streak, and five black spots near the base. The caterpillar is said to be green, with white dots, and to feed on a species of dock. It is found principally in Wales and Huntingdon-shire. There are three other species of this pretty genus, but they are so extremely rare that it would be useless to describe them in this elementary work ; for this reason I shall also be satisfied with merely naming the beautiful

MANCIPIUM DAPLIDICE, *Bath White*, a very rare insect, which received its English name from a piece of needle-work executed at Bath by a young lady, from a specimen of this insect said to be taken near that place.

PARNASSIA APOLLO, the *Apollo Butterfly*, is a doubtful native, but so very beautiful that we are tempted to wish it were less uncommon.

SPHINX.

Generic Distinctions.—Antennæ rather long, and increasing in thickness nearly to the apex, which is slender and hooked ; *proboscis* very long, slender, and convoluted (rolled up); *body* long and not tufted ; *palpi* three-jointed.

SPHINX LIGUSTRI. (Plate XII.) *Privet Hawk-Moth.* This fine insect sometimes measures upwards of four inches ; the upper wings are ash-grey, slightly tinged with rose-colour, and marked with black veins ; the hinder portion brown ; the margin having two white lines, which unite near the apex, on a greyish ground; the surface of the inferior wings is rose-colour, traversed by three black bands, that next the base being short, and at right angles with the others, which are parallel with the hinder margin ; the thorax is dark-brown and white at the sides ; the body, deep rose-colour, with black bands, broken in the middle by a broad longitudinal brown stripe, having a black line down the middle. The caterpillar is very beautiful, being of a fine apple-green, with seven oblique purple and white stripes on each side; the horn is yellow and black; it generally feeds on the privet.   The moth is not unfrequent in the south, and is seen occasionally in Scotland.   There is another very rare but beautiful species, SPHINX PINASTRI.

HEPIALUS.

Generic Distinctions.—Antennæ very short, sometimes slightly serrated; *palpi* and *maxillæ* both wanting.

HEPIALUS SYLVANUS. *Orange Swift.* This species is orange-colour, with three white lines ; the upper wings also variegated with chestnut-colour, and a small dusky spot on the disc.

HEPIALUS HUMULI, *Ghost Moth*, is pure satiny white, margined with yellow ; the female being entirely of that colour, with streaks of brown on the anterior wings. In this genus the female is unlike the male in colour and markings ; the caterpillars feed on the roots of plants, and, previous to their change, they bury themselves in an oval cell, composed of grains of sand held together by silken threads ; the eggs resemble grains of gunpowder. The *H. humuli* is sometimes found in June. All the species of this beautiful genus are very local in their habits.

GASTROPACHA.

Generic Distinctions.—Antennæ short, curved, and bipectinated in both sexes; *palpi* very hairy; *body* very large in the female; *wings* strongly dentated; when in repose, the anterior edge of the upper pair projects considerably above the lower, giving an oval form to the outline, and making it appear somewhat like a serrated leaf.

GASTROPACHA QUERCIFOLIA. (Frontispiece.) *Lappet Moth.* Surface of the wings rusty-brown colour, varying in shade, the extremity rather of a violet hue; the upper pair with three oblique, waved, blackish lines, and a black spot in the centre; the hinder wings are generally unspotted, though sometimes marked with faint streaks like those on the upper pair. The caterpillar is of a large size, the prevailing colour being dusky grey, with two blue spots near the head, circled with black; each segment has a fleshy appendage at the sides, and a tubercle on the last joint; it feeds on the willow, bramble, etc. The moth is found in many parts of England, but seldom in profusion.

LEPIDOPTERA. BOMBYCIDÆ.

ODONESTIS.

Generic Distinctions.—*Antennæ* bipectinated; *palpi* long, and projecting like a beak; *wings* entire; *body* long, the male having a tuft of hairs at the extremity.

Plate XIII.

1. Sphinx Convolvuli 2 Cossus ligniperda. 3 Tinea ?...

ODONESTIS POTATORIA. (Frontispiece.) *Drinker Moth.* In this species, the male is reddish-brown, the anterior wings having a patch of ochre-yellow at the base, and the disc slightly suffused with that colour; a rust-coloured line extends from the tip to the inner edge; there is also a faint line near the base, and another towards the hinder margin; on the disc are two white spots, one of them stained with yellow; the inferior wings are unspotted, but there is an indistinct transverse streak on each darker than the surface; the female is of a pale ochre-yellow, sometimes nearly white. The caterpillar has two long conical tufts, and on each side a series of black velvety spots, followed by a line of yellow, and, beneath these, tufts of white hairs. Like many of the tribe, it rolls itself into a ring when alarmed; it feeds on a variety of grasses, and is plentiful in many situations.

LEPIDOPTERA. HEPIALIDÆ.

COSSUS.

Generic Distinctions.—*Antennæ* short, and pectinated in the inner edge; *palpi* distinct, and three-jointed; *wings*, upper pair much larger than the under.

COSSUS LIGNIPERDA. (Plate XIII.) *Goat Moth.* The

wings measure between three and four inches ; the colour
of the upper pair is ashy-white, clouded with brown, and
marked with numerous black streaks, waved, and frequently
crossing each other; the hinder wings are brown, with faint
streaks near the hinder margin.   The caterpillar is very
large, of a lurid red, tinged with pale yellow, and having
a patch of chestnut-colour on each segment ; it does not
consume the foliage of trees, but derives its nourishment
from the wood, making its way through the stem, thus
doing much injury to the trees.   It lives three years in the
larva state, and then scoops a hollow in the wood, which it
lines with a warm fabric composed of the raspings of wood
and layers of silk.   Within this nest the chrysalis remains
till the moth is matured, when the apparently difficult task
awaits it, of coming to the surface, as it is impossible for
the moth to emerge and develop its wings in the confined
cell; this is accomplished by means of a series of small spines
projecting from the hinder segments, which, when one side is
moved forwards, prevents its sliding back, and the other is
brought up in the same manner.   The caterpillar diffuses a
very peculiar odour, which is said to resemble that of the
goat ; it is much more common than the perfect insect.

In spite of this powerful scent, strong enough to betray

the presence of the larva in the trees to which it resorts, and which cannot be supposed to render it very inviting, this caterpillar was fattened with flour by the luxurious Romans, and considered by them a great delicacy. Mr. Kirby, in the following passage, seems to recommend this example to our notice. " No insects," he says, " are more numerous than the caterpillars of *Lepidoptera;* if these could be used in the stock of food in times of scarcity, it might serve the double purpose of ridding us of a nuisance and relieving the public pressure." Réaumur also suggests this mode of diminishing caterpillars, and says, that if we took to eating them, he should be of the same mind as the red-breasts, and eat only the naked ones. I have elsewhere mentioned the different tribes of people who make these little creatures their food.

CHAPTER IX.

SEPTEMBER.

HOWEVER exhilarating and delightful may be the bright frosty mornings of this month to the sportsman and the pedestrian, however numerous the attractions which nature still spreads around for her true admirers, the Entomologist begins to find a great diminution in his peculiar sphere of action ; for though many insects still bask lovingly in the warmth of the noonday sun, enjoying to the last its enlivening rays, there are now seen comparatively but a small number of these airy creatures, who depend on warmth and brightness for their short career; and the butterfly-net, though by no means useless, will not be in such constant requisition as during former months.   There are, however, found at this season two of the most beautiful species of

Lepidoptera, which may be in all probability added to the collection,—the *Vanessa Atalanta* or Red Admiral, whose lovely wings are seen quivering with intense happiness, as it settles on a dahlia or other autumnal flower, or sometimes winging its way through an open vista of a wood, looking rather like the produce of the tropics than an inhabitant of our northern clime; and the *Acherontia Atropos,* or Death's-head Moth, which is the largest lepidopterous insect found in England, and, with one exception, the largest in Europe, as it sometimes measures nearly five inches across the wings, and females have been found of even still larger dimensions. The somewhat formidable English name given to this insect is derived from a large yellowish mark on the thorax, bearing considerable resemblance to a human skull, or "Death's-head." The great size of this insect, the peculiar appearance on its thorax, and the power it possesses of making a mournful sound when alarmed, render the Death's-head Moth an object of great curiosity and interest. There are many conjectures as to the mode by which it emits the cry referred to : some authors suppose it to proceed from the friction of one organ against another, such as the head and thorax, or the tongue and palpi ; others imagine it to be caused by the motion of the wings ; but more recent investigation has led to the belief that the sound proceeds from the interior of the head, where there appears to be an organ fit for the purpose. The caterpillar, when alarmed, has likewise the power of making a rather loud noise, like that caused by an electric spark.   These circumstances have produced a dislike and apprehension of this innocent and beautiful insect, which it will now be superfluous to assure my readers are perfectly groundless ; though Réaumur tells us of the members of a convent being thrown into the utmost consternation at the appearance of one, which happened to fly in during the evening at one of the dormitory windows.

Bees have a just cause for fear when they see one of these Moths enter their dwelling, as they have a great predilection for honey, and will disperse the inhabitants of the hive to obtain it, notwithstanding their numbers and their stings. Huber, who first noticed this circumstance, seems to be of opinion that the Bees are paralysed by fear, either from the great size of the intruder, from the sound it emits, or some other influence ; he also states that he saw the Bees in one hive, as if expecting their enemy, barricade themselves by means of a wall of wax, which completely obstructed the entrance, but was penetrated by passages for one or two workers at a time, thus securing themselves by an admirable sagacity against an enemy they could not resist. " The art of war amongst Bees," he says, " is, therefore, not restricted to attacking their enemies ; they know also how to construct ramparts as shelter for their enterprises ; from the part of simple soldiers, they pass to engineers," —another interesting proof of the instinct so abundantly possessed by these little creatures. Compare this rational fear and wise precaution with the feelings excited by the appearance of this Moth in Poland : the account is taken from the " Journal of a Naturalist :"—" The insect is called the ' Death's-head Phantom,' the ' Wandering Death-bird,' etc. ; the markings on its back represent to these fertile imaginations the head of a perfect skeleton, with limb-bones crossed beneath : its cry becomes the voice of anguish, the moaning of a child, the signal of grief ; it is regarded not as the creation of a benevolent Being, but the device of evil spirits (spirits enemies to man), and fabricated in the dark ; the very shining of its eyes is thought to represent the fiery element whence it is supposed to have proceeded. Flying into their rooms in the evening, it at times extinguishes the light, foretelling war, pestilence, hunger, and death, to man and beast."

In spite of this alarming account, the best wish I can give the young naturalist is, that one of these beautiful creatures may come within his net, that he may have an opportunity of admiring its magnificent size and singular colouring, and of adding a somewhat rare insect to his collection.

COLEOPTERA. HELOPHORIDÆ.

HELOPHORUS.

Generic Distinctions.—*Antennæ* of nine joints, and clubbed; *palpi* moderately long, the last joint thickened and oval; *body* oblong and depressed; *legs* not ciliated (hairy), though the beetle is aquatic.

HELOPHORUS AQUATICUS is about a quarter of an inch in length, of a dull brassy-brown, with the elytra greyish. This family comprises a few genera of minute aquatic beetles, which seem to form a connecting link between the true water and land-beetles; they creep very slowly, and may often be observed in muddy water, and on aquatic plants; sometimes in fine weather they leave the water and take flight; in some genera the palpi are very long.

GRYLLUS.

Generic Distinctions.—Antennæ long and setaceous, formed of many indistinct joints; *palpi* four; *head* large; *thorax* compressed; *elytra* inclined; *legs* formed for leaping.

GRYLLUS VIRIDISSIMUS. (Plate III.) *Great Green Grasshopper.* Body and elytra of a fine green, the former about two inches in length, the latter still longer, and comparatively narrow; the antennæ are also of great length. These insects are found at the beginning of autumn in grassy places, and in hedges by the sides of woods; they fill the meadows with their singing, but become silent on being approached. They are herbivorous, feeding in all their states on grass and herbs; though it is said that when in confinement they will devour each other. The long ovipositor with which the females are provided enables them to deposit their eggs at a considerable depth in the earth in small cells. The young ones when hatched resemble their parents in form and activity, but they are destitute of wings and elytra; in the state preceding that of the perfect insect, they have the rudiments of these organs. In this genus the eyes are very large, and the front of the head is acute and projecting.

R

VANESSA.

Generic Distinctions.—See page 129.

VANESSA ATALANTA. (Plate VII.) *Red Admiral,* or *the Admirable.* This insect is common in most parts of England, and in Scotland, where it is said to appear as early as July. The surface is a deep glossy black, the superior wings having a broad band of red running from the anterior margin obliquely across the surface nearly to the hinder edge, where it curves inwards ; beyond, there are six white spots and a faint blue stripe ; the under wings have a wide red border, on which there is a series of small black spots, and two patches of blue at the inner angle ; on the under side of the anterior wings the band is ochreous-red ; towards the base of the wings there is a waved streak of blue, and two others of red, one of which is united to the central band ; beyond the latter are two blue stripes, and the tip of the wings is tawny, with two small white spots ; the hinder wings are marked with undulating lines and spots, composed of brown, black, and grey, with a yellowish tinge. The caterpillar is greenish-black, with a line of yellow on each side, and feeds on the nettle, preferring the seed to the

leaves; it usually protects itself from the weather by drawing a few leaves round it, securing them by silken threads.

LEPIDOPTERA. SPHINGIDÆ.

ACHERONTIA.

Generic Distinctions.—*Antennæ* and *proboscis* very short, the former terminating in a hook; *wings* not indented.

ACHERONTIA ATROPOS. (Plate XI.) *Death's-head Hawk-Moth.* The primary wings dark brown, marked with several waved stripes of deep black and red, the latter colour marking the nervures on the hinder margin; near the centre of the wings there is a round whitish spot; the secondary wings are deep yellow, with two dark bands; the head and thorax are similar in colour to the upper wings, the latter having a large spot somewhat resembling a human skull. The caterpillar is sometimes five inches long, of a fine yellow, with seven oblique bands of green on each side, and a series of blue spots on the back, which is also marked with black; it feeds principally on the potato, and is much more common than the perfect insect, as many of them die before completing their transformation. These fine insects are found in various localities both in England and Scotland, and appear to have been very abundant two years since.

R 2

LEPIDOPTERA. SPHINGIDÆ.

SPHINX.

Generic Distinctions.—See page 230.

SPHINX CONVOLVULI. (Plate XIII.) *Unicorn Hawk-Moth.* This fine species is of a greyish ash-colour; the upper wings clouded with brown and black, also marked with lines of deep black; the secondary wings have three dark bands; the thorax is grey, with two indistinct lines in a horse-shoe form, having a black spot behind, and a red one adjoining; the body is marked with black and red in alternate rings, divided down the middle by a broad grey stripe, with a black line in the centre; there are two dark spots on the under side of the body; the wings frequently measure four inches and a half. The caterpillar is most commonly of a bright green, with black spots on the back, and oblique yellow stripes on the sides; it feeds on the convolvulus, and is not very common.

LEPIDOPTERA. ALUCITIDÆ.

ALUCITA.

Generic Distinctions.—*Antennæ* rather short; *palpi* long and slender, the terminal joint very long, acute, and turned upwards; *thorax* not crested; *body* short.

ALUCITA HEXADACTYLA. (Plate XIV.) *Many-plumed Moth*. This pretty insect is often seen in autumn creeping on the inside of windows, and is easily distinguished by the beautiful structure of the wings, which are divided into feathers, each having a central shaft fringed on both sides; the anterior wings have each eight of these rays, the hinder ones only four; the body measures about half an inch; it is of an ashy-grey colour, slightly varied with brown and white; a small black spot is visible on the tip of all the plumes, and the anterior wings have two brown bands edged with white.

LEPIDOPTERA. ALUCITIDÆ.

PTEROPHORUS.

Generic Distinctions.—*Antennæ* rather short; *body* long and slender; *legs* very long and delicate; *wings* divided into separate plumes, the upper from two to six, the under into three.

PTEROPHORUS PENTADACTYLUS. (Plate XIV.) *White-plumed Moth*. This is the largest species of the genus, the wings sometimes measuring more than an inch, the anterior deeply cleft; the whole insect is snowy white, with a silken gloss; the eyes are black. The caterpillar is white, tinged with green, and having a yellow line on each side; it feeds on the nettle. The moth is common on hedge-banks and weedy lanes throughout England.

LEPIDOPTERA.   YPONOMEUTIDÆ.

GLYPHIPTERYX.

Generic Distinctions.—Antennæ as long as the wings; *proboscis* short; *palpi* longer than the head, somewhat curved; *thorax* not crested; *wings* nearly lanceolate, with very long fringe.

GLYPHIPTERYX LINNÆELLA. *Linnæus's Glyphipteryx.* Sometimes measures nearly half an inch; the anterior wings are orange, with round silvery spots rising above the surface; the base and apex of the wings black, with a metallic lustre; the fringe and the hinder wings are dusky. Sometimes this little insect is found in great abundance, but it is not very general.

CHAPTER X.

OCTOBER.

Our mutually pleasant labour of describing and searching for the various and beautiful insects found, even in England, by the industrious and patient entomologist, is now drawing to a close ; for though even Moths may be taken during the winter months, yet not many of the young students for whom this little work has been written, either could or would expose themselves at night to brave November's fog and December's ice for the chance of entrapping them. This month however still produces some specimens ; if the autumn be mild, the Red Admiral yet delights us with its brilliant colours and graceful flight ; and a few of the nocturnal tribe remain to be described, which are sufficiently hardy to set at defiance the approaching inclemency of the

season.   It has been mentioned before, that some Butterflies
exist through the winter ; specimens of the Peacock and
Tortoise-shell are seen taking flight on a sunny day in
March, or even earlier ; but if caught, they will invariably
prove very shabby specimens, giving unquestionable signs
of having survived the winter in some sequestered retreat,
instead of appearing fresh from the cocoon,

> " No longer reptile, but endowed with plumes,"

The Small Tortoise-shell has been found in the Isle of
Wight as early as the 8th of January, and it is supposed
that great numbers of this species pass the winter in a torpid
state; in Italy it is seen throughout the cold season.   Other
insects may also be observed through the winter.   Gilbert
White, in the " Natural History of Selborne," says: " Ivy
is the last flower that supports the Hymenopterous and
Dipterous insects; on sunny days, quite to November, they
swarm on trees covered with this plant, and when they dis-
appear, probably retire under the shelter of its leaves, concealing themselves between its fibres and the trees which it
entwines.   Gnats, Flies of several species, and some Moths,
are stirring at all times when winters are mild, and are of
great service to those soft-billed birds which never leave us.

On every sunny day the winter through, clouds of insects usually called gnats, appear sporting and dancing over the tops of the evergreens in the shrubbery. Hence it appears that these *Diptera* (which, by their sizes, seem to be of different species) are not subject to a torpid state in the winter, as many winged insects are: at night, in frosty weather, and when it rains or blows, they seem to retire into these trees."

Many of the insects which survive the winter, hybernate, that is, pass the cold season in a state of torpor, and are then ready, with the first warm sunny days in spring, to commence their active labours as founders of a new colony: this is the case with individuals in the Bee, Wasp, and Ant tribes. Even amongst Hive Bees it is supposed that six or seven-eighths are destroyed by the cold of the winter months, in spite of their sheltered habitation, and the greater proportion of the insect world are certainly produced fresh every year from eggs deposited by the parent immediately previous to her own death. With regard to the dwellings reared by this portion of created beings, enough has been said to prove that its members are no mean architects: the skill of the Wild Bees and Wasps, the mathematical nicety of the Hive Bee, and the galleried domes of the Ant, all entitle them to rank with the far-famed Beaver in architectural capability. On this subject mnch interesting and valuable information may be gained from a work devoted to its delineation, in which the reader will find fresh proofs of the instinct so largely bestowed on these little workmen.*

I cannot close this short, and necessarily imperfect account of the insect tribes, whose extreme beauty of form and colour entitles them to universal admiration, in a more suitable manner than with these beautiful lines, so descriptive of their habits and loveliness :—

> " Child of the sun, pursue thy rapturous flight,
> Mingling with her thou lov'st in fields of light ;
> And, where the flowers of Paradise unfold,
> Quaff fragrant nectar from their cups of gold ;
> There shall thy wings, rich as an evening sky,
> Expand and shut with silent ecstasy.
> Yet wert thou once a worm—a thing that crept
> On the bare earth, then wrought a tomb and slept.
> And such is man : soon from his cell of clay
> To burst a seraph in the blaze of day."

* Rennie, "Insect Architecture."

CALOCAMPA.

Generic Distinctions.—Antennæ setaceous; *proboscis* as long as the antennæ; *palpi* robust and scaly; *anterior wings* long and narrow, hinder margin dentated; *thorax* slightly crested; *body* depressed.

CALOCAMPA EXOLETA. *Large Sword-grass Moth.* This, which is the most common species of the genus, measures about two inches ; the prevailing colour is ochreous, in some parts inclining to reddish-brown ; the upper wings marked with dusky-brown, each having two ear-shaped spots near the middle ; the hinder wings are grey, with a dark spot towards the base; the thorax is dark-brown ; the body reddish-ochre, with bands of dark brown. The caterpillar is very beautiful, and of a rich green, the back adorned with two rows of white spots ; below this, is a yellow line, then a series of small round spots, and a red line above the legs ; it is a very general feeder, but seems to prefer lettuce and asparagus. The moth is by no means common.

LEPIDOPTERA. NOCTUIDÆ.

SCOLIOPTERYX.

Generic Distinctions.—Antennæ short, bipectinated in the male, slightly serrated in the female; *proboscis* rather short; *palpi* long and scaly, first joint short, the others rather long; *head* and *thorax* crested; *anterior wings* deeply cut, the other pair slightly dentated.

SCOLIOPTERYX LIBATRIX. *Herald Moth.* The expanse of the wings is about an inch and a half; the thorax and anterior wings are reddish-grey, with a red patch at the base, a round white spot on the disc, and two white transverse bands from the outer edge to the apex; the colour is grey, and the red patch is sprinkled with minute yellow dots; the under wings are brownish, darker towards the hinder margin. The caterpillar is slender, of a green colour, with a white line on each side. This beautiful Moth is very abundant in the south, frequenting those neighbourhoods where the willow and poplar grow, as the larvæ feed on those trees; it is sometimes seen in July, but is named the Herald Moth, from its appearing in profusion during October, thus heralding the approach of winter.

LEPIDOPTERA. PYRALIDÆ.

HYDROCAMPA.

Generic Distinctions.—Antennæ moderate, and simple in both sexes ; *proboscis* rather long; *palpi* four in number; *wings,* all the margins entire.

HYDROCAMPA NYMPHÆATA. *Beautiful China-mark.* This species measures nearly an inch ; the wings are white, the anterior pair with two brown stripes from the base to the middle, the remaining portion having two bands connected in the centre ; a narrow one forms the margin ; the posterior wings have also two brown transverse bands, the arrangement of which bands varies considerably. The caterpillar feeds on duckweed, and is not common.

LEPIDOPTERA. YPONOMEUTIDÆ.

ARGYROMIGES.

Generic Distinctions.—Antennæ as long as the wings; *palpi* short, filiform, and drooping; *wings* rolled round the body when at rest; *body* very slender.

ARGYROMIGES SYLVELLA. *Dark Porcelain.* This insect measures about three lines ; the anterior wings are white, with an ash-coloured band at the base, an angular one in the centre, and two others near the apex; all are edged with a dusky hue, and tinged with golden yellow : the hinder wings are white. The family *Yponomeutidæ* consists of numerous small moths, none exceeding an inch in size, and many being only a tenth part of these dimensions. It is distinguished from the family *Tineidæ* by having only two palpi, while the insects belonging to that group possess four. The *A. sylvella* has been principally found in the neighbourhood of London ; there is some uncertainty with regard to the time in which this little moth makes its appearance.

．

CHAPTER XI.

NOVEMBER.

THIS wintry month cannot be better employed by the young Entomologist than in arranging his cabinet, studying the habits and peculiarities of the insect world by means of more extensive works on the subject, and endeavouring to classify and name those specimens not mentioned in this little introduction to the subject, but which may have come under the notice of the industrious collector. I shall now give some brief instructions relative to the construction of a cabinet, for the benefit of those who may desire to form a collection ; and, first, as to the necessary instruments for capturing.

The entomological net is of various forms and sizes : the one I have used has a pole six feet long, to which is strongly fastened a small hoop of cane to support a green gauze net or bag ; by a dexterous turn of the hand, easily acquired by practice, Butterflies and other insects may be soon caught on the wing with this simple apparatus. For some of the *Lepidoptera*, such as the Purple Emperor, a much longer rod is necessary, but these will not probably come under the observation of the student until he has acquired some experience, when he will find ways and means for himself. The cane, or hoop, may be about two feet in circumference, and the net half a yard in length. Entomological forceps are also very useful when insects are settled on leaves : these must of course be purchased, so that a description is unnecessary, and they may well be dispensed with for a time ; some chip boxes are also useful for putting insects in, when the collector is at a distance from home, that they may not be rubbed by the hand, and the colour injured. The *Lepidoptera* are most easily killed by a slight pressure of the finger and thumb on the thorax, below the wings ; this is, at least, generally sufficient, but sometimes, to the great uneasiness of the humane naturalist, they will be found alive some time after; suffocating by means of sulphur is still less certain, and Messrs. Kirby and Spence recommend the following plan :—

Fix in the lid of a small tin saucepan filled with boiling water, a tin tube consisting of two pieces, which fit into each other ; cover the mouth of the lower one with a piece of gauze, and place your insects upon it, then fix the upper one over it, covering the mouth also with gauze or muslin, and the steam from the boiling water will effectually kill the insects without injuring the plumage. Another more simple apparatus is a piece of elder, or any soft wood with the bark on, placed across the bottom of a mug ; stick the insect on this, inverting the mug in a deep basin, into which pour boiling water till it is covered, holding down the mug that it may not be overturned; in two minutes the insect will be dead." This latter plan, however, is objectionable, as the insect can hardly be fixed to the wood by any other means than by running the pin through the thorax when living ; and as only experienced naturalists can do this, with the certainty of deadening the feelings of the insect, and at once extinguishing the sense of pain, I would not recommend my readers to practise it. Others advise the head being touched with a strong acid, or that the insect be placed in a covered jar, half filled with bruised laurel-leaves; and beetles may be immediately killed by immersion in boiling water.

This is the only drawback to the study of this delightful

science in its perfection, namely, by forming a collection ; but it is very evident that insects have not any of the susceptibility to pain that is found in the higher order of animals, for they are frequently seen flying about with apparently the usual sense of enjoyment, when they have been deprived by accident of some of their due proportion of legs or wings ; I myself once caught a Tiger-Moth on the wing, which on examination was found to have lost nearly the whole of the lower part of the body.   The Rev. Mr. Bird gives a curious instance of the bluntness of sensation in insects ; he says : " When I was young in Entomology I wished anxiously to find the quickest mode of killing insects, and having captured a pretty beetle, I took a pair of scissors and divided it at the junction of the thorax and trunk ; the parts fell on a piece of white paper which lay before me.   Far from being dead, I was grieved and surprised to see the head and fore legs begin to run about the paper ; it occasionally stumbled, but rose again, and exhibited, if I may so speak, perfect self-possession ; it made for the edge of the paper, but arriving there, and looking down, it seemed to think it too precipitous, and so coasted along in search of an easier descent, which it did not seem able to find.   This searching for a con-

venient place of descent, suited to its curtailed condition
with respect to legs, of which it appeared perfectly aware,
occupied the head incessantly. I regarded it with astonishment; here, then, said I to myself, lies the vitality of
an insect, the body is dead at any rate; but in this I was
quickly undeceived, for in about a minute after the body
had fallen on the paper, I saw the hind legs brought upwards, and employed in deliberately brushing and cleaning
the wing-cases; the legs were then withdrawn, the cases
raised up, the true wings expanded from beneath, and
all made ready for flight; but the body, seeming then to
become aware that there was no guide, the head, its former
companion, being in possession of the eyes, the design was
abandoned, the wings folded up in their usual beautiful
manner, and the attitude of rest again assumed. A more
perfect act of a sentient creature could not be exhibited;
the head continued to run about, and the body to clean and
expand its wings, the one for about twelve, the other sixteen hours, their energies gradually decaying, till they
appeared to perish, or rather to sleep."

From this and many similar anecdotes, it is apparent
that insects have not that acute sense of pain possessed by
the various tribes of animals, and it may assist in reconciling

the collector to taking the lives of those insects necessary to perfect his knowledge of their forms and habits ; for most assuredly no one who studies the history of this part of creation, and discovers the design for which each insect is formed, will ever wantonly destroy one of them, from weak prejudice or silly and cruel caprice. When the insect is killed, it must be prepared for the cabinet : a pin (short whites) should be run into the thorax, and forced sufficiently through to pierce the cork of the box or drawer, and hold the insect firm ; it should also be high enough to prevent the legs from touching the bottom. The best method of setting Lepidopterous insects is by having a small board made, with pieces of wood the same length, about three inches wide and half an inch in depth, glued on at intervals of about an inch, or something less ; then by sticking the pin with the insect into one of these intervals, the wings rest on the higher parts on each side, and can then be easily fixed with braces in the desired position. The brace is merely a piece of cardboard, long and narrow, fixed with pins into the wood, and bearing gently over the wings so as to keep them steady in their natural attitude; the antennæ and legs should also be extended and kept in their places by pins. When quite stiff and firm, the insects may be removed to their proper place in the cabinet. The best method of arranging insects is in columns, with the generic name at the head of each column, and the specific affixed beneath the insect. A different drawer or box for each order is desirable for the young student, but not indispensable : and these arrangements must depend on means, and increase with his increasing collection. Two of each insect, particularly in *Lepidoptera*, should be procured, if possible, to exhibit both the upper and under side. Should an insect become mouldy, a brush dipped in spirits of wine in which a little camphor is dissolved, will clean it; but the insect must be dried before being replaced in the cabinet. Camphor in the drawers is very necessary to prevent the attacks of other insects, whose presence is easily discovered from the dust under the specimens; when this is seen, they should be taken out, brushed with a camel's hair pencil, then touched with spirits of wine, and placed near the fire till dry. Moths frequently change colour from an oily matter common to all insects; when this is the case, powder some dry chalk on a heated iron, cover it with a fine piece of linen, and apply to it the under part of the insect; the heated iron dissolves the greasy substances, which the chalk then absorbs. Insects, when stiff before being properly

placed out, which will sometimes be the case when caught at
a distance from home, may be relaxed by being placed on
moist sand.

The pupæ are principally found during the first three
months of the year, in long grass, on trunks of trees, and
felled timber; also in April and May, on palings. Caterpillars may be discovered in great profusion in May. Butterflies abound in June, July, and August, Moths extend
into September. The young naturalist must bear in mind
that most, if not all, of the insects named in this little Introduction, may be seen in other months besides those in
which they are described, as their existence is by no means
confined to that period; and of many species there are
two or more broods in the year; but they will at all events
be found, in favourable situations, during or near the time
indicated. My readers must also remember that they will
meet with numerous insects not even alluded to here; but
this will not cause surprise, when it is recollected that
in the order *Diptera* alone, there are nearly 2,000 species
in this country, and still more in that of *Coleoptera:*
but as the order in which any insect ought to be placed
will be readily known by those who have paid attention
to their peculiarities, reference may then be made to those

works entering into generic and specific details which it is evident must be omitted here. The best method of proceeding will be to secure a few specimens in each order, so as to be thoroughly acquainted with these great divisions, and then to fix on one or two of the most attractive, for instance, the *Lepidoptera* and *Coleoptera*, adding to the collection, and gaining all the information possible of these interesting groups, and by degrees the sphere of study may be enlarged according to the taste and opportunities of the student.

CHAPTER XII.

DECEMBER.

A SHORT account of the devastation caused by some insects
to our trees, fruit, etc., with the benefits we derive from others,
may not be uninteresting during this month of comparative
inactivity.   Many Coleopterous larvæ feed on the seeds of
plants: one species of Weevil, *Balaninus nucum,* destroys the
nut; *Bruchus granarius* lives in the pea, and all the species
of the genus inhabit seeds ; the genus *Apion* has the same
taste.   *Apion frumentarium* is known in the larvæ state as
the red Corn-worm ; it consumes the farinaceous portion
of the corn, whilst leaving the husk untouched, thus
causing great destruction in granaries.   A long list might
be made of *Curculionidæ*, which are injurious to seeds
of different kinds ; and the genera *Balaninus* and *Antho-*

nomus attack nuts and stone-fruits. Among the other orders, it is more particularly the *Lepidoptera* and *Diptera* which in the larva state live on fruit; the plum, apple, and pear are attacked by a small Moth, *Tinea Pomona;* another infests the chestnut; and the larva of the Moth *Pyralis fasciata* lives in grapes, destroying the largest and most beautiful fruit. Many of the Dipterous maggots distort plants by the excrescences they raise; others attack the ripening ears of corn; one of these produces the Fly called by Kirby *Tipula tritici.* The roots of plants furnish a supply of food to other species. The Coleopterous genus *Elater* attacks the roots of corn; one, whose larva is called the Wire-worm, frequently does much mischief; the thick yellowish larvæ of the *Melolonthidæ* and *Cetonidæ,* which in the beetle-state feed on leaves, devour in their imperfect form the roots of plants. The *Melolontha vulgaris* is very voracious; Dr. Burmeister mentions an instance in which the fields belonging to a farmer near Norfolk were entirely destroyed, and eighty bushels of the larvæ were collected. Radishes, carrots, and onions are all infested with the maggots of various Flies; as well as the cauliflower and cabbage. Trees are also much injured by the larvæ of Beetles, instances having occurred of whole pine-forests in

Germany being destroyed by their attacks ; and the tropical palms afford nourishment to many large exotic Beetles. The Dipterous and Lepidopterous larvæ take also an active part in destroying the roots and stems of plants and trees. Leaf-devourers are very numerous; among Beetles the *Chrysomelidæ*, both in the perfect and imperfect state, choose leaves as their food. Burmeister mentions the curious fact, that Beetles and their larvæ never consume the leaf from the margin, like the caterpillars of the *Lepidoptera*, but bite a hole in the centre, round which they eat: thus these destroyers may be distinguished merely by the appearance of the leaf on which they have been feasting. The Hymenopterous insects are many of them leaf-eaters, and 'the family *Tenthredinidæ* (Saw-flies) sometimes destroy whole plantations; but by no order are leaves so universally used for food, as among the *Lepidoptera*. The oak feeds numerous caterpillars ; according to Roesel, this tree supports 200 distinct species, and every forest-tree is supposed to nourish many varieties ; most of our fruit-trees also feed their peculiar tenants. Grasshoppers will eat almost every kind of leaf, but give the preference to grass. The devastation of the Locust is too well known to need description, and happily for us we are exempt from its ravages.

Numerous insects find their food in animal matter; the *Ichneumonidæ*, as described elsewhere, select almost exclusively the caterpillars of *Lepidoptera* as food for their young, thus preventing their two great increase. The parasitic insects infest birds and Mammalia; the genus *Pulex* (the Flea) lives principally on the latter; the *Hippobosca equina* on the horse; and the *Œstridæ*, in their larva state, on different quadrupeds, such as the ox and deer. But if many animals are annoyed by being made the food and habitation of insects, and even man suffers, both in his person, food, and clothing, by their attacks, how much is nature in general benefited by their unremitting labours in destroying noisome and decayed substances, and lessening the two great redundancy of both animal and vegetable life. There are also many more obvious uses: to the smaller animals, and numerous birds, they present an acceptable food; many birds feed exclusively on them, and the majority of bats, the shrew, the hedgehog, and the mole prefer them to other nourishment.

The advantages derived by man from the insect world are very numerous; many have been already named, and we are also indebted to it for some portion of our food and clothing. The *Cossus* of the Greeks and Romans was the

larva of a Beetle, and considered a great luxury; Grasshoppers are said to furnish nourishment to the Bedouins;
Locusts are eaten in Africa, and the Brazilian tribes are
very fond of a species of Ant which has an agreeable acid
flavour.   The honey of the Bee is mentioned as food even
in the earliest times, and is almost universally used at the
present day.   Some insects are used medicinally; the
Beetle employed in making blisters is one of the genus
Cantharis, or *Lytta*, according to some authors, and other
insects have been used for the same purpose; the *Coccinella*
(Lady-bird) is sometimes applied for toothache.   Formic
acid, obtained from Ants (*Formicidæ*), was formerly employed as a volatile stimulant; and galls, produced by the
puncture of many small Hymenopterous insects, have also
their medicinal properties.

Our clothing is indebted to the Silkworm, whose cocoon
furnishes the raw silk from which so many useful and
beautiful articles are manufactured.   The genus *Coccus*
presents us with a brilliant red dye; and we are indebted
to the gall-nut for one of the ingredients in ink: the true
gall-nut proceeds from the *Cynips Gallæ-tinctoria*, which
is found in Asia Minor.

It may interest some of my readers to have a slight idea

of the insects existing in remote ages, now that the science of Geology has gained so distinguished a place in the studies of every lover of nature. They are chiefly found imbedded, according to Dr. Burmeister, from whose valuable work I abridge this notice, in a resin called amber, which is cast on the shores of the Baltic, or found in the more recent strata. The way in which insects have been enclosed in this sub- stance can be no other, than that they stuck to it when in a fluid state, and were enclosed by what continued to exude from the tree; according to the rapidity with which this took place, depends the condition of the enclosed insect, those which were quickly enveloped being in the most per- fect state of preservation. The species of insects found in amber seem to differ but little from those now existing, and the number of different species already discovered is con- siderable, showing that the class of insects must have been then, as now, the most numerous: they are chiefly those families which exist in woods and trees. Of the Order *Coleoptera* are mentioned several of the families *Elateridæ*, *Chrysomelidæ*, and *Curculionidæ;* of *Hymenoptera, Ichneu- mons*, a small kind of Bee, and numerous Ants are specified. Amber *Lepidoptera* are very rare; *Diptera*, on the con- trary, are common in the families *Muscidæ, Tabanidæ*, etc.

Tipulidæ are uncommon ; the Order *Neuroptera* presents specimens of the family *Phryganidæ*, and a species of *Hemerobius*, with a few others. Amongst *Dictyoptera* have been observed individuals of the genus *Ephemera*. In *Orthoptera* the *Blattidæ* are the most numerous, some small *Achetidæ*, a few larger Grasshoppers, and Locusts.

Fossil insects have been also discovered in recent formations, of which Marcel de Serres has given the most complete list. According to him, they were found in calcareous marl, which separates the several strata of gypsum in the quarries of Aix in Provence ; they are accompanied by impressions of plants, and are chiefly those insects which live in a sandy soil ; their colour is gone, as they are of a uniform brown or black. The list includes specimens of the Beetle, Ichneumon, Ant, Butterfly (very rare), Fly, Dragonfly, Earwig, Grasshopper, Cricket, and many others whose names are less familiar : these will suffice to give the geological student a slight idea of the insect world in ages long since passed away.

These, and many other particulars given in this little volume, are not intended to satisfy, but to excite the curiosity of the young naturalist, and induce him to apply for information to other sources ; but it may be hoped that they whose research extends no further, will not rank with those who overlook the beautiful works of God through culpable indifference or ignorance. " We attach, and with reason," says Réaumur, " a kind of consequence to the knowledge of the faults and perfections of the productions in the fine arts, such as music, poetry, painting, sculpture, and architecture ; but of the works of the Lord of Nature, of the Master of Masters, we scarcely think. There can, indeed, be no room for criticism where there is nothing but what is admirable, and where the most perfect finite intelligences, the more they study such subjects, the more they discover of their wonders. Yet this knowledge, so well calculated to elevate the mind, and lead it to the contemplation of the source from which all these wonders proceed, is regarded by many as frivolous, or of little importance. But he who looks upon an insect as merely a particle of putrid matter, and who has no idea of the marvellous organs of these minute animals, is in a state of ignorance far more blamable than the man who should confound the most finished productions in the Fine Arts with the most rude and shapeless masses." The student of nature will never rise from his labours disappointed at imperfection and incompleteness in the object of his pursuit ; the most minute, the most unwearied investigation, only leads to the discovery of increased perfection, and must, if duly considered and rightly directed, strengthen every feeling of love and adoration towards the great Author of all things, exciting, in a high degree, the feeling expressed in those true and beautiful lines :—

> " What, though I trace each herb and flower
> That drinks the morning dew,
> Did I not own Jehovah's power,
> How vain were all I knew!"

ENGLISH

ALPHABETICAL INDEX.

T

275

LATIN

ALPHABETICAL INDEX.

REEVE'S
POPULAR NATURAL HISTORY.

In Square 16mo, with 20 Coloured Pages of Plates in each
Volume, containing from 80 to 200 Illustrations.

———✦———

1. *British Birds' Eggs.* LAISHLEY.
2. *British Crustacea.* ADAM WHITE.
3. *Greenhouse Botany.* CATLOW.
4. *Field Botany.* A. CATLOW.
5. *Geography of Plants.* DR. DAUBENY.
6. *British Mosses.* R. M. STARK.
7. *Palms.* DR. B. SEEMANN, F.L.S.
8. *British Seaweeds.* DR. LANDSBOROUGH.
9. *British Conchology.* G. B. SOWERBY.
10. *British Ornithology.* GOSSE.
11. *Mammalia.* ADAM WHITE.
12. *Mineralogy.* H. SOWERBY.
13. *The Aquarium.* G. B. SOWERBY, F.L.S.
14. *Mollusca.* MARY ROBERTS.
15. *Garden Botany.* A. CATLOW.
16. *Economic Botany.* ARCHER.
17. *British Ferns.* T. MOORE.
18. *British Lichens.* LINDSAY.
19. *Physical Geology.* J. B. JUKES.
20. *Zoophytes.* DR. LANDSBOROUGH.
21. *British Entomology.* M. E. CATLOW.
22. *Birds.* ADAM WHITE.
23. *Scripture Zoology.* M. E. CATLOW.
24. *The Woodlands.* MARY ROBERTS.

LONDON: ROUTLEDGE, WARNE, & ROUTLEDGE.

In 1 vol. price **2s.** cloth boards.

RURAL ECONOMY FOR COTTAGE FARMERS
AND GARDENERS. A Treasury of Information on Cowkeeping, Sheep, Pigs, Poultry, the Horse, Pony, Ass, Goat, the Honey Bee, Farm and Garden Plants, &c. By MARTIN DOYLE, and others.

In 1 vol. fcap 8vo, price **2s. 6d.** cloth, extra gilt.

THE KITCHEN AND FLOWER GARDEN ; or, the
Culture, in the Open Ground, of Bulbous, Tuberous, Fibrous-rooted, and Shrubby Flowers, as well as Roots, Vegetables, Herbs, and Fruits. With a Coloured Frontispiece. By E. S. DELAMER.

"A book that may be consulted with advantage by the practised gardener as well as by the novice in the art."—*Gardeners' Chronicle.*

In fcap 8vo, price **3s. 6d.** cloth gilt, or with gilt edges, **4s.**

WANDERINGS AMONG THE WILD FLOWERS:
How to See and How to Gather them. With Remarks on the Economical and Medicinal Uses of our Native Plants. By SPENCER THOMSON, M.D. A New Edition, entirely revised, with 171 Woodcuts, and Eight large Coloured Illustrations by Noel Humphreys.

"This book speaks for itself. No one can open it without being persuaded of its value, and without obtaining information that will cultivate the mind and improve the taste. Both as respects matter and illustration, it is an admirable production.—*Bell's Weekly Messenger.*

In post 8vo, price **6s.** cloth gilt, or **6s. 6d.** gilt edges.

A NATURAL HISTORY. By the Rev. J. G. WOOD.
The Second and Cheaper Edition, with many additions. Containing nearly 500 Illustrations, from original designs by William Harvey, engraved by the Brothers Dalziel. The book is printed on tinted paper, and its principal features are :—

1st. Its Accuracy. 2nd. Its Systematic Arrangement. 3rd. Illustrations executed expressly for the Work. And 4th. New and Authentic Anecdotes.

"One of the most recent and best of Messrs. Routledge and Co.'s publications." —*Times.*

In post 8vo, price **6s.** cloth gilt.

ENGLISH COUNTRY LIFE. Containing Descriptions of Country Life, of every Form, Manner, and Custom. Sketches of Ploughmen, Sowers, Shepherds, Milkmaids, Haymakers, Reapers, Gleaners, Swineherds, Foresters, Fishermen, Hunters, Gipsies. Earth-stoppers, &c. &c. Wood Paintings of Wide Marshes, Windy Wolds, Wild Woods, Ancient Forests, Parks, Chases, Thorpes, Granges, &c. &c. Picturesque Roadside Resting-places, Country Churchyards, Villages, Ferries, and Footpaths. Flowers of Spring, Summer, Autumn, and Winter. Descriptions of British Birds, Animals, Reptiles, Insects, and Fishes. Old and Modern May-day Games, Harvest Homes, and Christmas Merry-makings. Rural Sports—Nutting, Blackberrying, Squirrel Hunting, Fox Hunting, Stag Hunting, Hare Coursing, Shooting, and Fishing. Riverside and Seaside Scenery ; with Pictures of Sweet Spring-time, Long-leayed Summer, Golden Autumn, and Gloomy Winter. By THOMAS MILLER, Author of "A Day in the Woods," "Gideon Giles," &c. With nearly 300 Illustrations by Birket Foster, Gilbert, W. Harvey, &c.

In one vol. crown 8vo, price **5s.** half-bound.

ILLUSTRATED BOOK OF DOMESTIC POULTRY. Edited by MARTIN DOYLE. With Twenty Illustrations, from Designs by Weigall. Printed in Colours.

This Work also includes the Treatment of Turkeys, Geese, Ducks, Guinea Fowl, &c. ; and to show the practical nature of the Work, the following Contents are given :—The History of the Origin of Domestic Poultry—The Process of Incubation and Progressive States of the Egg —The Selection of Stock for Breeding—Methods of Hatching, Rearing, and Feeding Chickens—The Egyptian Artificial Mode of Hatching, and our Modern Experiments—The best Modes of Feeding, and the Result of Experiments with various Kinds of Grain—Places for Poultry-houses, their Position, and the proper way of Housing Poultry—Methods of Fattening and Caponizing—The Organs of Digestion explained, and the Diseases to which Poultry are liable, with Advice for their Treatment, as well as for Accidental Injury.

To all poultry keepers this book will be invaluable, as it will save them in time its price every year. It is the most useful, practical, and complete book that has ever been issued on domestic poultry.

www.ingramcontent.com/pod-product-compliance
Lightning Source LLC
Chambersburg PA
CBHW060522030726
47498CB00004B/1036